Astronomy Saves the World

Securing our future through exploration and education

Daniel Batcheldor

Spacewalk Publishing
Merritt Island, FL

© 2016 Daniel Batcheldor
All Rights Reserved
http://danielbatcheldor.com/

Jacket design by Davinica Nemtzow and Madeline Heckman

Published by Spacewalk Publishing
http://spacewalkpublishing.us/

Printed in the USA by Four Color Print Group, Louisville, Kentucky

First Edition

ISBN 978-0-9972475-0-3

For Emily and Charlotte

Never stop asking why.
Don't trust people who claim they have an answer to everything.
They don't.
If someone can't tell you something without it making sense, it probably doesn't make sense.

Table of Contents

Preface 1

PART ONE

THE PROBLEMS

Chapter 1: The Science Struggles, the Human Troubles 7
    The Science Factor 13
    The Human Factor 17

Chapter 2: Avoidable Problems 29
    The Last Mass Extinction 30
    Forces, Energy, and Momentum 31
    Impact of Impacts 37
    Comets 41
    Water 44
    Finding the Valuable Ones 50

Chapter 3: Unavoidable Problems 57
    The Fate of the Universe 57
    Nuclear Energy 60
    Typical Stars 66
    Pulsating Stars 69
    Death of Stars 70
    Death of Dead Stars 72
    Fate of Our Sun 74

PART TWO

THE PRINCIPLES

Chapter 4: Astronomers Are Really Good at Staring 79
    Some Simple Questions 81
    Some Mathematics 84
    Earth Is Not Flat 91
    The Closest Things Are Far Away 95

Chapter 5: The Scale of the Problem 103
    There's No Harm in Checking 104
    More Mathematics 106

Accuracy and Precision 108
Parallax 110
The Wandering Stars 115
The Distance to the Sun 124
The Distance to the Nearest Stars 128

Chapter 6: Space and Time 131
Stepping up the Distance Scale 132
Importance of Brightness 134
Standard Candles 139
Astronomical Rainbows 142
Our Expanding Universe 145
Age of Our Universe 148
Age of Earth 152

Chapter 7: Fair and Balanced 159
Our Universe 160
Galaxies 162
The Motions of Stars 167
Remnants of Stars 172

PART THREE

THE FUTURE

Chapter 8: Rock to Rock Exploration 177
The Ultimate Destination 180
Getting There 183
Moon Base Alpha 185
Asteroid Mining 189
The Mars Problem 192
Beyond Mars 199
Getting off the Ground 203

Chapter 9: Emancipation via Education 205
Intellectual Armor 207
The Fight for Knowledge 212
Impact of Knowledge 215
Astronomy Saves the World 220

Acknowledgements 225

Glossary 227

Persons of Note 237

Notes 241

Astronomy Saves the World

Preface

As with any idea, when I started to put together the material for this book, I immediately thought, *Someone must have already done this.* Well, if someone has, I can't find it. But that doesn't mean I've read all the astronomy books or all the comments ever made about the value of astronomy.

This is not the first time that the phrase "Astronomy Saves the World" has been used for the title of something. First, there is a brief 1993 article by Mark Bailey, called "Applied Astronomy Saves the World," which argues for observing programs to spot potentially hazardous near-Earth objects[1]. Second, Ed Lu (astronaut and co-founder/CEO of the asteroid-hunting B612 Foundation) included the phrase in the title of a lecture he gave to the University of Hawaii Institute for Astronomy in August 2013. "Astronomy Saves the World: Protecting the Planet from Asteroid Impacts" speaks straight to the heart of the obvious role astronomy has in dealing with the potential threats of asteroids[2]. While that is discussed somewhat in this book, asteroid deflection is only a small role that astronomy can play in saving the world. There is a much bigger story here. A story I think is worth telling, and a story for everyone.

I've tried to keep scientific jargon to a minimum and rely on only basic algebra skills for the mathematical explanations. I hope readers of many ages and backgrounds will find at least some of what is presented useful, interesting, or thought provoking.

I am an astrophysicist, which means I have a broad, but fairly shallow, knowledge across many areas of physics. This is one of the drawbacks, or, as some may argue, the delights, of being an astrophysicist. We are the jack-of-all-trades physicists that need to understand enough of relativity and quantum mechanics to investigate the complete history of our origins. This is not to say astrophysicists aren't experts at anything. When we need to, we delve deeply into one very narrow field of astrophysics, rummage around until we reach some kind of limit (usually imposed by underdeveloped technology), and then go on to find something else. On the other hand, some astrophysicists delve even more deeply into even narrower fields and can spend their entire careers in them.

The knowledge needed to appreciate our universe plays to one of the key points of this book, and it has somewhat determined the level I have tried to adopt; no one needs a degree in astrophysics to understand the basic facts about our universe.

Throughout the book, I use the phrase "our universe" rather than "the universe." This may seem confusing at first, considering that the textbook definition of "universe" is *all existing matter and space*. But the multiverse conjecture is one way to explain why our universe is the way it is. It is possible that other universes—in which matter and space don't play such obvious roles—exist.

Astronomy is chock-full of lazy metaphors, and I will preemptively apologize for that. These metaphors are a classic consequence of living in a counterintuitive universe. Some things in astronomy are either weird or incomprehensible in everyday experiences, and so comparative examples may be given that aren't entirely accurate but try to reflect the idea at hand in a less mind-bending way.

In many cases, I purposefully don't back away from the grassroots explanations needed to understand some astronomy, but there are some other things that have been presented in a fairly superficial manner. This is either because I know my own limits in understanding the entire subject or because going further would compromise the aim of not needing a degree in astrophysics to understand what is presented here. However, there are definitely some things included that are intellectually challenging. A great way to know whether we understand a concept is to see if we can coherently explain or teach it to another person.

If you are a scientist, engineer, or enthusiast reading this, you may not learn anything new about physics and may find at least a few explanations frustrating. However, I hope there are other parts that you find useful. The something new could just be a different way of explaining something, or it could be a new perspective on a topic you already understand. It could just be the realization that you don't, and can't, know everything or go through life without making mistakes.

Most of what I have written is based on my own experiences as an academic: the classes I have taken, the classes I have taught, and the astrophysical research I have done, including the huge amount of background reading that goes with it. Nevertheless, despite double- and triple-checks, genuine errors could have crept in, and you may spot one. Perhaps someday there will be a second edition with your correction acknowledged. If you find something I've written that is factually incorrect, by all means, please politely point it out to me with the appropriate references. I simply hope some of you enjoy what I've written and perhaps—if you are still in the position to do so—consider pursuing one of the many careers that can be used to help save the world.

Astronomy Saves the World

Part One

The Problems

Astronomy Saves the World

Chapter 1: The Science Struggles, the Human Troubles

Undoubtedly, our universe is a truly astonishing place. For millennia, science has struggled against human ignorance to determine this. Despite the cognitive troubles, we've known the shape of Earth and the distance to the moon for thousands of years. We know the distance to the sun and the staggering distances to other stars. We know that Earth has been orbiting the sun for billions of years, that our constituent atoms were forged in the hearts of long-dead stars, and that our universe is nearly fourteen billion years old. We know Earth is a target for devastating impacts with comets and asteroids and our sun will sterilize our planet in the distant future. These are all truly profound facts, but how do we know them, and how many should we understand? Would it make any difference in our lives if all of us understood them? I think it would. These facts guide us toward our origins and propel us toward our future. They humble us, inspire us, enlighten us, and unite us.

On our journey to discover the truths of our universe, our species has relied on the science we've been doing since becoming human: astronomy. Since we first turned to the stars and contemplated the celestial mysteries offered by the night sky, astronomy has provoked in us a sense of wonder and curiosity.

Since then, astronomy has become quite complex and has parented the more comprehensive field of astrophysics.

A common question is, "What is the difference between astronomy and astrophysics?" Well, besides astrophysics not being frequently mistaken for the unavailing exercise of astrology, the difference is apparent in the names themselves. An astronomer can provide us with basic data about an astronomical object, while an astrophysicist uses physics and mathematics to calculate the finer details. Today, the majority of professional astronomers are really astrophysicists who rely heavily on science, technology, engineering, and mathematics (STEM). Without these things, we could not have a comprehensive understanding of what we observe in our universe, including ourselves!

If an astronomer and astrophysicist looked at the same star, they would likely give different sets of information. An astronomer would know where it is in the sky, its approximate distance, and its temperature. An astrophysicist would take things further and calculate the star's approximate age, what it will turn into when its fuel runs out (white dwarf, neutron star, or black hole), and—among many other things—its relative abundance of elements like hydrogen, helium, carbon, nitrogen, and oxygen. Understanding the origin of elements like these is important for a number of reasons, not the least of which is the fact that everything important to us is made from them.

Just as astronomy and astrophysics are specializations in physics itself, we find there are additional divisions within these subjects: cosmology, galactic astrophysics, stellar astrophysics, and planetary science, all of which go a long way in helping us have a complete understanding of our cosmic origins. However, planetary science, which is the study of non-stellar bodies in our own solar system and other star systems, has a notable difference when compared to "mainstream" astronomical research. Planetary scientists are able to send spacecraft to the objects they study, whereas the classical astronomer cannot.

Astronomy Saves the World

A recent success in planetary science has been the *New Horizons*[1] mission. The pictures from this spacecraft, which zipped past Pluto in the summer of 2015, are jaw dropping. They have revealed a staggering number of profoundly interesting facts about one of the best-loved members of our solar system, and they will help piece together its history. However, planetary science has a much wider role to play. It has helped us understand the influence of various human activities on Earth's atmosphere by studying the atmospheres of Mars and Venus[2]. In addition, planetary science is driving the possibility for humans to successfully survive on, and exploit the intrinsic properties of, astronomical bodies other than Earth, such as the moon, Mars, and asteroids. Clearly, planetary science is currently one of the most important branches within astronomy, and it will continue to play a key role in understanding our past, our present, and our future.

If there are other scientific fields that are important for saving the world, then why not call this book *Science Saves the World*, *Physics Saves the World*, *Mathematics Saves the World*, *Planetary Science Saves the World*, or simply *Education Saves the World*? Well, in my experience, people feel a certain allure toward astronomy. I have yet to come across anyone who has said, "Yeah, so I saw the stars the other night, and they were crap." Typically, the response to seeing a dark sky on a cloudless night is more like, "Wow! Look at that! I want to know more!" Compared to the other sciences, astronomy does have a huge advantage in getting this response, which also takes very little effort to provoke. If the weather is favorable, we can simply step outside at night, throw our hands in the air, and shout, "Ta-da!" Physicists and chemists typically need to go to some lengths to get the "wow" response; they basically must make something explode. In this case, the response is less like, "I want to know more," and more like, "Do it again!" One needs only to sit through a few episodes of *MythBusters* to see that point demonstrated [3].

Astronomy Saves the World

Regardless of why we find the stars so effortlessly compelling, astronomy has matured to the point where we understand our universe is a remarkably complex, but not completely inexplicable, place. In fact, as technology advances, we are able to gather increasingly breathtaking data on astronomical objects. What used to be cloudy, dim smudges on the night sky are now glorious star-forming nebulae, the death throes of stars, or other galaxies at distances the human brain struggles to grasp. The more details we are able to see in these objects, the more complex and esoteric they become.

The same effect happens when we observe the building blocks of these astronomical objects at sub-atomic scales. Here, we are left with a disturbingly confusing and wickedly counterintuitive world of quantized energies, wave functions, and probabilities— the intricacies of this quantum mechanics are still being uncovered. Having to use the physics of the incredibly small to describe the properties of the incredibly large is one of the many things that provoke our senses of wonder and curiosity. It also makes astronomy a sneaky, educational honey trap. It draws unsuspecting people into the reasoned world of STEM that is advancing our culture toward a more positive future.

Much like the night sky, the future leads many of us into the deepest of wonders. This is where we rest our fears. This is where we lay our hopes. But many of these fears can be overcome, and many of these hopes can be met, by realizing our ambitions in space.

Before we became a space-faring species, there was no way to prevent the types of terrifying cosmic impact with comets or asteroids that could wipe out our species. Now, we are regularly sending objects into space. We have shown that we can rendezvous with these asteroids and comets. We have also had some limited success in landing on them (the *Hayabusa* mission to Asteroid 25143 Itokawa and the *Rosetta* mission to Comet 67P/Churyumov-Gerasimenko [4,5]). If needed, we have the ability to nudge these things away from Earth. All we need is a little more effort. Catastrophic cosmic impacts should be a thing of the past.

With our ever-increasing robotic treks around our solar system, there is also now hope that some record of the human race will march on even if we completely fail and the whole planet is destroyed. With the help of one of astronomy's most respected champions, Carl Sagan (1934–1996), long-term preservation of some of our culture has already occurred. Sagan helped convince NASA to inscribe information about ourselves (messages to potential aliens) on plaques attached to *Pioneer 10* and *11*, and he later helped decide the data inscribed on gold records stowed on *Voyager 1* and *2*, which are now traveling in interstellar space[6,7]. But regardless of the uncrewed technologies we fling at the stars, to truly ensure the longevity of our species, we must permanently get enough people off this slightly damp, rock-covered metal ball we currently all call home.

The reasoning here is quite simple: Considering what the long-term astronomical future has in store, we face extinction if we idly remain on Earth. No matter what we do, Earth will become uninhabitable; we can't save it. As this is the case, then why not call this book *Astronomy will Save Civilization*? In my experience, people tend to disassociate themselves from the word civilization. It insinuates some long-forgotten ancient culture rather than the societal construct of which we are all currently part. People don't think about the civilization in which they live, but they do care about the world in which they live. So while "the world" is used throughout, it is really our civilization that astronomy will save.

Regardless of the nomenclature, when (not if) our universe presents us with a scenario that jeopardizes our planet, if we have not made the right choices to try to prevent that scenario from happening, then clearly astronomy will not save the world. Should the worst happen and a massive rock hits our planet, it will likely not be the fault of the astronomers anyway. It will simply be a result of the necessary financial resources not being made available when they were needed the most.

We already know what the problems are and how to solve those that may occur sooner rather than later. This, of course, assumes we don't succumb to our own idiocy and, inadvertently or purposefully, annihilate one another without any help from our universe. But should we manage to avoid such self-annihilation, there is no guarantee our global society would progress toward a more prosperous future. We could slip back into previous levels of barbarism or succumb to new ones. History reassures us that things are generally progressing toward a more halcyonic global community (some of us have walked on the moon, after all), but past societal trends offer no security. Nevertheless, astronomy clearly has had, and continues to have, an ongoing role in our saving ourselves. It has formed the backbone of many scientific and philosophical discussions and revolutions, most notably putting us in our cosmic place time after time [8]. It has shown our species that we occupy no special place in our universe. This compels us to consider our own fragile existence and has instilled some of us with a modicum of cosmic humility.

If astronomy truly is to save the world, it must be capable of doing several things. First, it must provide a complete and accurate picture of the mechanisms that pose a clear and present danger to our planet. Second, it must provide realistic solutions to those dangerous mechanisms. Third, it must provide a way for our species to overcome our differences and unite for the greater good, i.e., make us provide the brightest possible future for the next generations. Finally, it must provide universal emancipation via education so everyone can become aware of, and concerned for, the challenges we may face and will have to overcome together. Only then can everyone have the opportunity to live happy lives free from tyranny, totalitarianism, and fear. This is a fairly tall order. It is not just a matter of training people on a list of facts. It is also a matter of training people to apply the critical thinking approaches needed when confronted by new claims or evidence.

As will be discussed and demonstrated in the following chapters, astronomy gives us the opportunity to know the array of astronomical problems we face and shows us the solutions to these problems. It also shows that these solutions are not necessarily difficult; they simply require time, money, and determination. Astronomy will also show us our path forward as our species migrates to the stars. Finally, astronomy provides an easy, and generally compelling, gateway to education, and can unite humanity with a common interest that transcends social, political, and theological differences. Not only can astronomy save us from our universe, it can save us from ourselves. It is a powerful subject.

The Science Factor

Science is the only method we have for uncovering the physical problems our universe poses. But the more we have learned about our universe, the more we have been able to manipulate it to our benefit. Scientists from all fields of study play a fundamentally important, and often overlooked, role in helping to improve our world. Some of our work provides spin-off technologies, like medical diagnostic equipment and solar cells, which have a genuine and direct benefit to the public as a whole. We don't just gawk at pretty pictures, attend conferences in exotic places, and mess around with instruments costing billions of dollars. We also uncover remarkable facts about our own place in space. We exist on the surface of a planet that is about four and a half billion years old. Our species has likely only existed for the last several hundred thousand years. Earth is not the center of our universe but in orbit around a rather unremarkable star, in a rather grand galaxy, in a rather lonely region of a staggeringly beautiful universe—a universe that has a demonstrable beginning around fourteen billion years ago. We have also uncovered some very alarming facts about what is in store for the tiny parts of Earth currently habitable by humans. Two of the most serious of these astronomical facts are impacts with asteroids and comets, and the ongoing outbursts from, and evolutionary future of, the sun. The inevitability of these events becomes clear when considering the staggeringly long periods of time on which our universe operates. Such periods of time are difficult to imagine and hard to understand.

Despite all the things our universe could throw at us in order to end the world, we may also consider all the things that threaten our future that seemingly have nothing to do with astronomy. Providing for an increasing global population, climate change, sustainability, pandemics, nuclear war, or a technology blackout are just some of the self-inflicted disaster scenarios we face. However, even in these cases, our experiences with astronomy may have played a role in leading us to contemplate these potentially catastrophic outcomes. It is interesting to note, for example, that the Center for the Study of Existential Risk was co-founded by Astronomer Royal Martin Rees[9].

Professional astronomers and astrophysicists must be experienced critical thinkers and complex problem solvers, because we are working to find answers to mostly unique scenarios based on limited amounts of data. This gives us the ideal skillsets to undertake world-saving research; it's not just a matter of *what* we think but *how* we think. Astronomy can therefore be used to encourage more critical thinkers, of any age, to become interested in any of the sciences that could help prevent many of the presently conceived, and those yet to be conceived, doomsday scenarios. The mysteries of our universe grab their attention and can show them what is possible.

When we move past our individual instinct for self-fulfillment and preservation and consider ourselves one species struggling to avoid extinction rather than one person trying to get rich and avoid death, we are thrown into a commonality that does not divide us based on our differences. To put this into perspective, it is a fact that we all live on the same cosmically insignificant planet among billions of others. Earth is small, fragile, and just happens to have a tiny fraction of its volume capable of hosting an oasis of complex life. We are all so incredibly fragile that if we moved just seven miles up from where we are now, the shift in atmospheric conditions would cause us to pass out in roughly thirty seconds and die shortly after[10].

The facts about our place in this universe seem to make some people uncomfortable, and many go to great lengths to either ignore or insulate themselves against them. Many people seem to do this by *believing* in weird things. However, astronomers embrace the facts of our universe, which are based on the application of logical arguments applied to reproducible evidence. This in and of itself makes terms such as *believe* and *belief* incorrect when used in reference to scientific findings. These terms also insinuate that scientists, and the *scientific method*, operate on some level of *faith*. This is absolutely not the case. The scientific method is not a system of belief. It is how we work out what is correct and what is incorrect. It ensures scientists report all the possible explanations based on the empirical data available and known, and sometimes new, scientific principles. It ensures we whittle down explanations until there is only one that is entirely consistent with all the data. Many people seem to struggle to understand this about science.

A singular explanation is not always the case, however, and we frequently find there are two or more explanations consistent with collected data. In these cases, the explanations are considered further, and tests are designed to rule out one or the other. Occasionally, even new tests cannot rule out an explanation, and the most plausible (usually the simplest) is preferred, and the principle of Occam's razor, which was named after William of Ockham (1285–1347), is applied. If future observations provide additional data, scientists will use the new data to update the preferred explanation. Scientists shamelessly change their minds on a regular basis, because this action is at the heart of the scientific method. Unfortunately, the public often considers changing one's mind a sign of weakness, and we see this when discussing "flip-flopping" politicians. This skewed belief often leads to some hesitation when it comes to trusting the results of scientific studies. This is another struggle science must face.

The handling of the results from the BICEP2 experiment (the second experiment to attempt **B**ackground **I**maging of **C**osmic **E**xtragalactic **P**olarization) makes for a good demonstration of the scientific method, and perhaps a not-so-good demonstration of how to present new findings to the public. To explain why our universe looks the way it does, very early in its history, there must have been a period of exponential inflationary expansion. Such a violent event would have caused gravitational waves (ripples in the fabric of our universe) and thus a predictable polarized signal in the cosmic microwave background, i.e., the cooling fireball left over from the event that began our universe.

In March 2014, the BICEP2 team announced they had observed a signal consistent with those predicted by this inflationary model[11]. If confirmed, this would have represented a huge step forward in our ability to probe the state of our universe just a tiny fraction of a moment after it began. It would have also provided observational evidence of this key inflationary hypothesis in cosmology. However, the BICEP2 team did caveat their result, explaining that the signal could be the result of dust in our own galaxy. Things got a little dicey when the team presented their interpretation of the data to the media before independent peers had reviewed their paper[12]. This is a scientific faux pas. In short order, another team also pointed out the dust problem, and then both teams worked together to determine whether the suspected signal, which would have indicated inflation had taken place, was actually viable. One interpretation was right; one was wrong.

Both teams accepted the results, and the scientific community now favors the dust explanation[13]. The scientific method worked, and we are still waiting to detect that new predicted signal to support inflation.

The BICEP2 misstep highlights an interesting discussion about what is more productive for the public impression of science. On the one hand, we have a demonstration of the self-correcting nature of the scientific method, and on the other hand, we have given the public the impression that scientists do not know what they are doing. Could this latter impression be fanning the flames of the current anti-intellectual movements? Should scientists conduct the scientific method behind the scenes until they establish the preferred explanation, and thereby maintain a mysterious—but credible—impression? Or, should the public be shown the scientific method so they can appreciate its value or naively cast it aside as indecisive? Ideally, the public should be trained in the scientific method and shown that to naively cast it aside is foolish. Without the knowledge of its power, science will always struggle to unequivocally convince the public of its findings.

The Human Factor

Our universe is so despairingly huge, and timescales so long, that incredibly violent, and seemingly rare, astronomical events occur regularly. As humans generally have trouble understanding the scales of space and time, this is one of the many counterintuitive things about our universe. Stars die every day, just usually not near us. Large rocks hit planets every day, just usually not Earth. However, if we wait long enough, one of these events will occur close enough to Earth to end us all. This fact might be somewhat more concerning if our lives weren't so desperately short. Since they are, we rarely think about it. Astronomy probably won't physically save the lives of anyone, either. When that becomes necessary, the honor will likely go to modern medicine. In fact, improved living standards and advancements in modern medicine have already helped the average world life expectancy more than double since 1900, going from a little over thirty years to around seventy years[14]. But at some point in the future, after our generations are but faded memories, astronomy will save the world. It must. Our species has a natural instinct for survival, and to let the world end would be to go against a core component that makes us human. Failure is not an option[15].

As a scientist, I must caveat almost anything I say and find that arguing with myself is a worthwhile exercise; it makes for good practice in defending or promoting ideas and facts in public. To demonstrate how this works—and to continue the current thread of morbidity—it can be noted, for example, that for those of us lucky enough to avoid all the other causes, time could be the reason for our death. Since time is an inherent property of our universe, it might still be our universe that pulls our paltry plugs. We can go on to argue that as soon as we're born, we begin to die and that aging is a terminal disease. While birth certainly starts a countdown with an unknown end point, the cause of aging is still an ongoing subject of study. Starting somewhere in the mid-to-late twenties, aging essentially boils down to critical cells replacing themselves with ever-decreasing efficiency[16]. In time, our current cells are so degraded that they are no longer able to perform the various functions needed for our bodies to operate properly. It could then be argued that aging is curable, making the lack of a cure a shortcoming of medical science rather than an inevitable product of our universe.

Regardless of what stops our clocks (time, our universe, or medical shortcomings), we currently have a good enough understanding of our universe to be confident that there are no major astronomical threats that require immediate action. There are plenty of other problems we face, but large rocks on collision courses and exploding or expanding stars, for at least the next hundred years, aren't one of them. That is not to say that we shouldn't start to worry about them now, though. It is still a fact that if we do nothing, these threats will definitely become a major problem for future generations.

As incredibly terrifying as it is to think of our universe ultimately killing us, it doesn't actually have a destructive intent. It is not sentient. There is no genocidal cosmic master of puppets. It may *feel* like our universe is working against us with the everyday troubles we face, but most of these are simply a result of bad decisions, someone being an idiot, or a group of people making bad decisions or being idiots.

In some ways, it would likely provoke a greater response if our universe were trying to kill us. Perhaps it would get more attention, and more people would insist that we do something about it. Despite the obvious threats to the long-term survival of our species, we have not experienced a sudden increase in the number of astronomers—or everyday people for that matter—trying to save the world. This is something that obviously needs to be done, and not just in the comic book superhero sense, right? Therefore, we need to raise the stakes. To do this, we can take a dark cue from many of the movie bad guys: Don't go after a person, go after their kids. Instead of stating that our universe won't kill *us,* let's be sure everyone knows that if we do nothing, it *will* kill our descendants. There is no argument. If we sit here, idly twiddling our thumbs, sooner or later some astronomical event will end the world and take all our descendants with it. I suppose we could leave these issues to the next generation, but remember what I was just saying about being an idiot? Let's not be idiots; the time to do something is now.

The physical nature of our universe is not the only trouble our world faces. As previously mentioned, we have ourselves to contend with. This is clearly seen in the rather conspicuous sectarian differences that continue to cause significant loss of life every day. We also face apparent efforts to limit the progress of science, and therefore astronomy, from anti-intellectualism. This seemingly conservative phenomenon clearly has roots in the scriptures, and it continually attempts to undermine the education of many young children in countries all over the world. Alarmingly, such divisive methods may also be present at the highest levels of many governments. One need only to look at the attempts to defund reproductive health services in the United States and the dumbfounding behavior of the 2015 chair of U.S. Senate Committee on Environment and Public Works, who used a snowball to support the denial of global climate change[17].

Regardless of their motivations, groups perpetuating anti-intellectualism discourage our youth from entering into the science fields. This could leave future generations of scientists at a distinct disadvantage. Some significant worry comes from groups that attempt to pollute science textbooks used in the United States with equivocations[18]. Members of these movements may be manipulating the poor understanding of the scientific method and ignorance of facts to further influence public opinion, policymakers, and budgetary committees for their own benefits and beliefs.

Many of those under the barrage of anti-intellectuals will be the victims of recognized argumentative tactics that use flawed reasoning. These logical fallacies have been used both knowingly and unwittingly for millennia. The begging the question fallacy is one of the most popular. That is a circular argument that includes the answer to the argument that the proposer believes to be true. "How do you know that what is written in this book is true?" The circular argument is then, "It is true because it is written in this book!" Typically, this is used to convince others that some particular scripture has been personally delivered to the human race by a supernatural being. Of course, should one realize a passage in a scripture contains a logical fallacy, an apologist will argue that a different interpretation of that question is in some way flawed, while also ironically maintaining that their particular interpretation is not flawed.

It would be very beneficial to the progress of humanity if the general public were well versed in "The Fine Art of Baloney Detection," as Carl Sagan put it in a chapter from his book, *The Demon-Haunted World: Science as a Candle in the Dark*[19]. But detecting baloney—the essence of the logical fallacies—shouldn't have to be a fine art. Indeed, the phrase itself (perhaps purposefully done) is almost an oxymoron, because there is no fine art to baloney.

Being able to recognize and identify each and every possible logical fallacy during the course of a conversation is a skill most of us don't likely have the resources to master. It is also something science itself does not directly provide a comprehensive shield against. So, a simple approach to spotting when someone may be using a logical fallacy is to wait until that person "crosses the line." By doing so, that person has stepped over the normal boundaries of civil discussion and has probably rendered their own argument ridiculous in the process.

What constitutes a reasonable boundary doesn't necessarily need to be anything unusual, and the approaches used in everyday conversation are perfectly fine. In this case, the end of a civil conversation is signaled by *a lie, a threat,* or *a personal insult.* Think of these as pocket-sized baloney detectors.

Of these three things, lies—or more strictly false or inaccurate information—are probably the most difficult to deal with. In some cases, lies are used with the express purpose of exploiting people, usually with the intent of collecting money and/or gaining power. In other cases, a person is simply unwittingly passing on false information. Some people do this with glee if that information supports a previously held belief, opinion, or bias. This is a case of ignorance rather than malice, but the information being spread is still false and will completely discredit and destroy the argument that person is trying to make. The trick is to pick out false information and remain skeptical of any new or extraordinary claims.

For something to be considered true, it must have at least two strengths. It must provide reproducible physical evidence in support of the claim and pass a falsifiable test. Basically, for something to be considered true, it must be proved beyond all reasonable doubt and make a prediction that can be observed. This is partly how a properly administered and civilized court is supposed to operate. This is how science operates. This is also the perfect way for someone to exploit the burden of proof logical fallacy. In this case, rather than someone having to prove their own claim, someone else is challenged to disprove it. Frequently the impossible task of proving a negative is given, and the inevitable failure is used as evidence that the original supposition, however ridiculous, is true. "You can't prove there isn't an invisible unicorn dancing on the moon; therefore, there must be an invisible unicorn dancing on the moon." A modification of this can be seen when someone is challenged to thoroughly explain a complex area of science that contradicts a belief-based claim. An unsatisfactory explanation is then used as evidence that the science in question is flawed. This is termed the "fallacy fallacy," and while it would be ideal if everyone could satisfactorily describe the origin of our universe and evolution on the spot, it is technically unnecessary because of the wealth of peer-reviewed literature to support these realities in the first place.

In the interest of fairness, it must be noted that making the peer-reviewed literature defense can be construed as a logical fallacy, because it is an appeal to scientific authority. This is a fairly weak criticism, as the peer-review process is the best way we currently have to support the scientific method. An author of a study must convince other experts in their field that their work has merit and has followed the scientific method. But, if you'd rather avoid getting caught by this argument, it is probably worth becoming familiar with the evidence for climate change, a non-eternal universe, an old, non-flat Earth, a set of common ancestral species, the benefit of vaccines, and humanity's visits to the moon. In many cases, a basic knowledge of astronomy will help you understand these facts. At least then, the next time you come across a person claiming that evolution is a lie or that NASA faked the moon landing, you can be ready with your well-reasoned rebuttal or a simple facepalm. Unfortunately, if that person doesn't value evidence and reason over their own beliefs, and is not prepared to change their mind, then what reasonable evidence or argument could be presented in order for them to do so? The trouble with many humans is that they don't listen, they aren't very reasonable, and they don't like being told they are wrong. Good scientists must listen, they can only be reasonable, and they are told they are wrong all the time.

If you find yourself caught up in these types of discussions, do not underestimate what you do and do not know, even if you are not a scientist, a science major, or an aspiring scientist. Simply be reassured that there are still things that we don't yet understand, and that's okay. And while these things may turn out to have complicated naturalistic explanations, there are some very basic facts about our universe that are easily understood. Why do kites fly? Why is the sky blue? What is evolution? How can you prove Earth is a geoid and staggeringly more than six thousand years old? What is the evidence for the Big Bang beginning to our universe? If you can't explain these things, you shouldn't be surprised if tricksters try to take advantage of your own ignorance and hound you with their own flawed worldviews.

In the event that lying or using logical fallacies doesn't work on you, you may find yourself confronted with threats. Threats can be real or imaginary, and people commonly consider them to be a sign of a weak argument. They can force a point without proper explanation. As threats are intended to intimidate, they can be spotted easily. It should be obvious when someone has threatened to hurt or murder another person in a hideous fashion and have them damned to be posthumously tortured for eternity in an imaginary location. In these cases, phrases such as, "You are going to burn in hell for thinking that" or "I'm going to kill you because you said something I don't like," are usually thrown around, and all reasonable discourse has come to an end.

The typical direction of threats, from aggressor to defendant, can be turned around. This is a tactic anyone could use, right or wrong, at any time in order to try to shut down an argument or emotionally trick the naive into relenting. One can simply state a feeling of being threatened, or that a safe space has been invaded, even if no threat has been made. This may cause the aggressor to back down, even if the argument they are making is valid. What seems increasingly common in the current social climate is that some people appear very eager to claim offense in the face of reason, as if this type of "cry-bully" victim complex, encouraged by so-called social justice warriors, somehow constitutes a valid point in the conversation. It does not. It's baloney. Similar behavior is seen when someone claims to have been insulted.

Insults can be a little bit trickier to deal with. There will always be obvious and intentional insults (name calling or ad hominems), which attack a person's character in an attempt to discredit them. Other insults can be easily veiled behind philosophical points, which not everyone may have considered. Insults can be just as effective as threats, because the person posed with the question may shut down, especially if they have not yet learned that, for example, being good for goodness sake is more virtuous than being good for fear of punishment or in hope of some vague, and certainly not guaranteed, reward.

In the face of an actual insult, we are left with three possible outcomes: The insult is obvious and intentional, and all reasonable discourse is over. The insult is unintentional, and a rational rebuttal can be made to continue the conversation in good order. The insult is either intentional or unintentional, but rational thought is trodden on, so the conversation degenerates to threats.

Astronomy Saves the World

More often than not, outrage at being insulted, even if the insult is not intentional, is the end result. A good example is the claim of personal insult when someone has stated something that contradicts a person's ideology. Someone who truly believes Earth is flat should not claim to be insulted every time someone talks about Earth as a geoid, shows a picture of Earth from space, or draws a cartoon of a spherical Earth. Nor should it be okay for those pseudo-insulted people to claim they have the right to inflict some kind of punishment or take some kind of social or political action because they have had their feelings hurt. The disappointment continues when someone claims to be offended by a fact. "The Big Bang offends me!" Well, that's great, but it's not anyone else's problem. Taking offense does not offer anything to an argument, and pretending it does only helps perpetuate the idea that it should. This is ridiculous behavior. It's baloney.

As with feeling threatened, this claim of being insulted or offended will typically cause a person to back down even if the argument they are making is valid. There seems to have arisen some kind of overbearing social stigma attached to accidentally insulting or offending someone. So much so that even people who are making perfectly logical and valid arguments will silence themselves. Some groups are so successful at exploiting this stigma that people who would normally use reason to criticize manifestos and actions often end up defending them.

We must be very careful when discussing what causes offense. Being offended is just another unpleasant aspect of life. But when we limit some offensive person's right to speak freely, we immediately limit our own right to do so as well. How can we be sure that the things we say aren't going to offend somebody else? We can't. We shouldn't have to worry about it either. And because I am fortunate enough to live in a country that has laws to protect free speech, I'm free to speak my mind. Unfortunately, even in the twenty-first century, the citizens of many countries do not have this freedom [20].

Astronomy Saves the World

The same can be said for ideas. Yes, people are going to have some astoundingly poor ideas, such as women being subjugated to men and other such bigotry. Rightly, this nonsense deeply offends a huge number of people. But, until ideas like that emerge into actions and infringe on individuals' inalienable human rights, there is no way to step in without also stifling ideas that are genuinely genius. At the heart of these issues is the ability to freely and openly discuss and critique all ideas without having to suffer from lies, threats, and personal insults, which is something the sciences encourage. Today, astronomers have a good enough understanding of our universe to make all theological explanations about our origins completely unnecessary. We can also have reasonable discussions on a solid moral prescription, for the benefit of all humankind, without the need for supernatural supervision. The post-theistic era is emerging, but it hasn't been an easy journey. In the past, scientific advances have been purposefully hindered. This has made scientific and technological progress almost impossible to achieve, irrespective of the complexity of the subject at hand.

But our species had to start somewhere, and we can credit ancient religions as humanity's first attempts to explain the world. It is a shame that they quickly and lazily grasped at supernatural claims and beliefs rather than patiently waiting for the results of empirical testing. The fact that history is rife with examples of people endlessly killing and dying for mere beliefs is morally appalling. Or, perhaps many of us simply have a genetic predisposition for jumping to baseless conclusions and then defending them with our lives. It is obviously better to have a civil conversation and perhaps admit a mistake than to die fighting. Imagine where things could be right now if our ancestors had maintained a society that valued evidence over superstition.

Encouragingly, despite some people's best efforts to maintain a belief-based authority, it is clear that the general population, at least in developed countries and in the younger generations, is increasingly rejecting the explanations of the natural order offered by scriptures[21]. And, just like parents must get out of the way of their own children so they can reach their full potential in life, it is time for these belief systems to get out of the way so we—as a species—can reach our full potential.

The closer science gets to providing a set of complete, correct, and satisfactory answers to humanity's most compelling questions, the further from relevance the divisive doctrines are pushed. Contradictory belief-based worldviews, especially those that condemn many of their own members for being themselves, tend to do the opposite.

The direct contradictions to dogma that astronomy provides are not the only human factors this subject has had to overcome. There are academic ones, too. When starting out in astronomy, it doesn't take long for a student to realize this subject is well outside the normal experiences people go through every day. For example, a cursory inspection implies a flat Earth with the sun, moon, and stars passing overhead. Gravity is also not as straightforward as what meets the eyes. Earth seems to be pulling us toward it, but in actuality, both our mass and the mass of Earth are causing curvatures of spacetime that bring us together. Astronomy is not intuitive, and simply remembering a set of facts is not enough to understand the processes at play. Consider how most of us think about a sunrise or sunset. Do we think, *Wow, this is a beautiful sunrise*, or do we think, *Wow, I am standing on the surface of a relatively small ball, whose rotation is causing the sun to increase its altitude in the morning and decrease it in the afternoon*? Does our understanding of the latter make you appreciate the beauty more deeply?

The final hurdle astronomy has is essentially impossible to overcome. With the exception of planetary science, astronomy is a science where the system under scrutiny cannot be directly tested in order to gather more information. Compared to the other branches of the natural sciences, astronomy is one where we cannot change the environment in which we observe the various phenomena we are interested in understanding. This problem does not mean that astronomy itself is impossible; it just means we have our work cut out for us. Despite this challenge, we have determined our place in this universe without being able to poke, prod, probe, dig up, dissect, break, smash, dilute, mix, cool, heat, evaporate, detonate, clean, electrify, or magnetize any of the systems we study. We haven't even looked at the systems we study from the other side to see what shapes they really are. All we've basically done is patiently stare, and we've become really quite good at it.

Chapter 2: Avoidable Problems

Despite the scientific struggles and the human troubles, we have reached the point in the history of our solar system where one species holds the near-term fate of an entire planet in its own hands. Our scientific understanding and technological resourcefulness means that we can now avoid the problems of asteroid and comet impacts. Instead of allowing our home to be ruined again by a giant lump of rock (an asteroid), or a huge dirty snowball (a comet), we should now be able to manipulate the orbits of these objects. With the right actions, we can redirect any asteroids or comets with worrisome orbits and prevent any more catastrophic collisions with Earth. The dinosaurs had an excuse. We don't.

Understanding and preventing impacts from asteroids and comets is the obvious way in which astronomy can save the world. The damage they have caused in the past is in plain view all over Earth itself, and we see them regularly hitting Jupiter. Their destructive potential can also be determined by using some basic physics, and as long as we can find them all in the first place, this same basic physics allows us to figure out what is needed to redirect them.

Aside from their destructive potential, because of the water they contain, we must also consider asteroids and comets as resources that will help our species in the future. Water is going to be absolutely essential once humans start to migrate throughout our solar system and beyond, so the study of asteroids and comets is a fundamentally important area of astronomy for more than one reason.

The Last Mass Extinction

A massive object collided with Earth around sixty-six million years ago. A global mass extinction event followed. It is estimated that about three-quarters of all species then present perished, including that collection of terrible, overgrown lizards known as the dinosaurs[1,2,3]. So what is the evidence for this object? How big must that object have been? Are there any more on their way, and what can we do about them?

All over the world, there is a thin band of clay that contains unusually high levels of iridium. Physicist Luis Alvarez (1911–1988), already a Nobel Laureate, led the research team that discovered it in 1980[4]. This iridium layer is unusual for two reasons. First, it is found everywhere, meaning its formation must have been the result of an event that affected the whole planet. Second, outside of this layer, iridium is a very rare element. It is forty times rarer than gold[5]. Inside this layer, there is one hundred times more iridium than normally found in Earth's crusty surface layer.

Iridium, which is denser than iron, would have sunk to Earth's core when the planet was forming. Therefore, we expect the surface of Earth to be iridium poor and primordial lumps of the solar system that didn't become planets (asteroids) to have a fairly uniform distribution of iridium (as well as other metals).

Below this iridium-rich surface layer are rocks indicative of widespread, shallow, ancient warm seas[6]. In geological terms, these particular rocks come from a period in Earth's history known as the Cretaceous. While these rock formations contain the fossilized remains of dinosaurs, they abruptly end sixty-six million years ago at the iridium-rich layer.

Above this layer are different rocks. Here we find the fossilized remains of a growing number of mammalian species. This is known as the Paleogene period, or Pg for short, and these rocks have also been dated to around sixty-six million years. Together, the Cretaceous period and Paleogene period sandwich the iridium-rich layer, which is frequently referred to as the K–T or K–Pg boundary. The K refers to Cretaceous, and the T refers to Tertiary (which is a longer geological timescale of which the Paleogene is a part). So, we are looking for not only a mechanism that could distribute the iridium worldwide but also something that could have happened in an incredibly short period of time (compared to the geological timescales).

The evidence points toward an asteroid impact around sixty-six million years ago. This would explain the rich levels of iridium, the global distribution of the layer, the sudden change in climate, and the sudden change in the types of animals. The clinching piece of evidence for such an event was announced in 1991[7]. Radar and gravity mapping of Earth's surface revealed a crater "hiding" in the northern Yucatán Peninsula of Mexico near the small town of Chicxulub. The precise center of the impact was likely just a few kilometers offshore. The crater could be over twenty kilometers deep and over two hundred kilometers wide; it spans both land and ocean. It is one of the largest impact craters ever discovered, but what size asteroid could have made such a huge scar on the surface of Earth?

Forces, Energy, and Momentum

In order to estimate the size of the asteroid that impacted Earth, we must discuss energy. This will involve some simple physics first documented by Sir Isaac Newton (1642–1726). In 1687, Newton published his three laws of motion and his one law of gravitation in a collection of texts called the *Principia*[8]. The original Latin descriptions of Newton's three laws of motion sound weird as translated into eighteenth-century English by Andrew Motte, and I found this particularly distracting when having to learn them in high school[9].

1. Every body perseveres in its state of rest, or of uniform motion in a right line, unless it is compelled to change that state by forces impress'd thereon.
2. The alteration of motion is ever proportional to the motive force impress'd; and is made in the direction of the right line in which that force is impress'd.
3. To every Action there is always opposed an equal Reaction: or the mutual actions of two bodies upon each other are always equal, and directed to contrary parts.

Students readily remember the third law in the "for every action there is an equal and opposite reaction" form, but this wording does not encourage learning.

If we cut away the linguistic pomp, we find three simple facts about the motion of objects that we might consider common sense. First, if an object has a speed (including zero), it is going to keep on having that speed unless something pushes or pulls it. Second, if we push or pull on that object, the change in the motion will depend on the mass of the object; it is more difficult to move more massive objects. Third, when we push or pull on an object, that object will push or pull us back equally.

There are a number of things to bear in mind when considering Newton's three laws of motion: A push or a pull doesn't necessarily come from physical contact with the object. (The masses of two objects pull on each other because of gravity, and two magnets arranged north-to-north, or south-to-south, will push on each other.) If there is more than one push or pull, we must add them all together appropriately to figure out the total push or pull. Just because something isn't moving doesn't mean there aren't any pushes or pulls acting on it. (The pushes and pulls could all be balanced.) Just because something is moving doesn't mean there are pushes or pulls acting on it. If the velocity (the combined speed and direction of something) is changing, it doesn't mean that the size of the total push or pull is changing. Collectively, all these pushes and pulls are known simply as *forces*.

A lot of information has just been presented. It takes some effort to read, is easy to forget, and can be a bit confusing. What has been presented constitutes around 30 percent of an introductory college physics course. However, almost everything can be summarized with five symbols: $\Sigma F = ma$. The Σ symbol is the Greek capital letter sigma. It means carefully add up the symbols immediately following it. The symbol F is for *force,* and it represents the things that need to be carefully added up. These forces need to be carefully added, because they are *vectors*, and vectors (like velocity) act with some particular strength in some particular direction. It matters in which direction the forces point. We can put a book on a flat surface and push it with parallel fingers opposite each other on either side. If we push with equal and opposite forces, the book goes nowhere. Adding the two forces from our fingers results in zero. However, if we change the angle of one of our fingers, the two forces will no longer total to zero, they won't cancel each other, and the book may begin to move.

When ΣF doesn't equal zero, we must consider what happens to the object (mass m) given that a must now have some value. The a is for acceleration, and it, too, is a vector. To make sure we know it is a vector, the "a" is bolded and italicized, just like the F for force. Acceleration is the push backward we feel when we are in a vehicle that starts moving, or moves faster, and the pull of the seatbelt when we slow down. When the acceleration is positive, we speed up, but when it is negative, we slow down and eventually start going backward. When it is zero, we keep going at the same speed and in the same direction in which we had previously been moving (even if we were stationary). So, when an object isn't accelerating, i.e., it is stationary or moving without speeding up or slowing down, whatever forces are acting on that object must add up to zero.

The *m* is for mass. Note that mass and weight are not equivalent. Mass is just a number (a scalar quantity), whereas weight is a force (and therefore a vector quantity), because it is the mass times the acceleration due to gravity (a vector). For us on Earth, the direction of weight is always toward the center of Earth, because that is the direction of the acceleration vector.

All this talk about Newton's laws and forces is all very nice, but what does it have to do with energy, momentum, and a sixty-six-million-year-old asteroid impact? Consider a bowling ball and a basketball. Make both of them roll with the same velocity. Which one would you rather stop? Because our common sense says that the basketball will take less effort to stop, most people will choose that. The reason is that we know it is not *just* the velocity of something that determines how hard it is to stop. It is mass times velocity. Together, these two things are known as *momentum*. Something with more momentum (the bowling ball) is going to be harder to stop than something with less momentum (the basketball).

The question now becomes, "What *makes* it harder to stop?" We are attempting to change the velocity of the basketball from some number to zero. We are also trying to change the momentum of the basketball from some number to zero by applying a force (probably your foot or hand). Forces are responsible for changing momentum. If we apply a large force for a short amount of time, the ball will stop. If we apply a smaller force for a longer amount of time, the ball will also stop, but it won't be as sudden (think of a "soft" catch or "soft hands" when catching a cricket ball, baseball, or softball). We now also know it matters for how long the force is applied. This means that the force applied actually determines how *quickly* the momentum changes; there is an added element of time.

Let's move from stopping a bowling ball or basketball to pushing a car or bicycle. Nothing about the physics has changed. We will still be attempting to change the momentum of these objects by applying a force. Instead of going from some velocity to zero, we are now trying to get them to go from zero to some velocity. In addition, in moving both objects over one hundred meters, we are going to notice that one is going to make us more tired; we'll have less energy left, because we must work harder to push the car. The work we have done is the energy given to the car or bicycle, and more energy is given when the distance over which the force is applied is longer. The force applied over some distance is one definition of energy. Don't forget force is a vector though, so the direction in which the force is applied matters. We wouldn't want to push down on the roof of a car to get it to move along a road; we'd push it as parallel to the road as possible.

We can summarize the last few paragraphs, which now amount to about half of an introductory college physics course, as follows: Energy (E) = force (F) x distance (x). ***F*** = mass (m) x acceleration (***a***). Acceleration is velocity (v) / time (t). Velocity is x/t, i.e., velocity equals distance divided by time. As a consequence of these simple relationships, we see that energy is mass times the *velocity squared*. In words, it sounds horrible; the figure below may help.

The eagle-eyed reader may spot where momentum (mass times velocity) comes into play here, and that energy can also be described by the change in momentum (typically referred to as an "impulse") times distance. However, bear in mind that Figure 2.1 shows only the dimensional analysis and leaves out a lot of important details. For example, F→ma leaves out the details given by ΣF = ma. E→Fx also leaves out the cases when the force is getting larger or smaller and when there are two different directions between the total force and the displacement. The actual equations include calculus, vector products, and constants, like one-half.

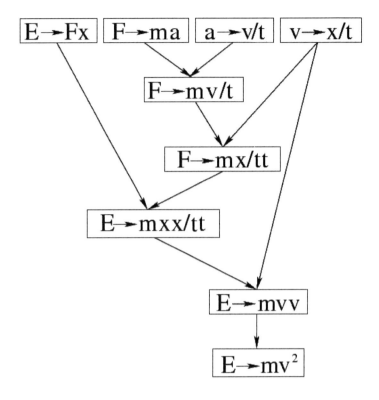

Figure 2.1: An example of dimensional analysis. In the top line, we have energy (E), force (F), acceleration (a), and velocity (v). x is distance, and t is time. The dimensions of a and v can be combined into F (second and third lines). The new dimensions of F can then be combined into E (fourth line). v is then combined into E (fifth line), and we find energy can have the dimension of mass times velocity squared.

The punch line is that the amount of energy needed to stop something dead is the same as its kinetic energy, and that is proportional to its mass times its velocity *squared*. Massive things that are moving quickly have a massive amount of kinetic energy. Light things that are moving slowly do not. Conversely, light things that are moving quickly (like bullets) can have a lot more kinetic energy than massive slow things. Taking the mass out of the equation makes things a little clearer. Consider two objects of the same mass, one moving twice as fast as the other. The faster one has *four times* the kinetic energy. However, if one were three times faster than the other, it would have *nine times* the kinetic energy. As relative velocities increase, energies quickly become terrifying.

One other thing that may be noticed from Figure 2.1 is that we have come across the form of the most famous equation in the world. As we now know that energy can be described using mass times velocity squared, and the fastest that anything can go is the speed of light (c), we find that Einstein's equation of $E = mc^2$ is staring back at us[10].

Impact of Impacts

A basic understanding of energy and momentum is all we need to crudely estimate the mass of that sixty-six-million-year-old Earth-impacting asteroid. The energy released in the impact will be equal to how quickly the momentum of the asteroid changed, or the kinetic energy it had just before it reached the surface. Regardless of which one we choose to use to make this estimate, we must remember that both momentum and kinetic energy depend on mass. If we can calculate the mass, we can make some assumptions about its density and shape to determine its size.

If a piece of space debris is large enough, it will leave a crater, and this is something we can physically measure. If the object doing the impacting isn't large enough to leave a crater, then we as a species don't really need to worry about it. Small pieces of solar system debris like this are not going to result in a catastrophic worldwide mass extinction event. Most objects burn up as meteors in the atmosphere due to heating from friction, and the remnants of some are found on the ground as meteorites. When a meteorite does reach the surface of Earth, very occasionally someone is unlucky (or lucky) enough to have some property damaged by it. This was the case in October 1992 when a twelve-kilogram meteorite hit Michelle Knapp's parked car[11].

The location of the largest known meteorite is about a five-hour drive north of Windhoek, the capital city of Namibia. It was named Hoba, because it was found on a farm called Hoba West. It is not only the largest meteorite ever found but also the largest known single lump of iron on Earth. Like the other iron on Earth, it would have originally formed at the core of a now-deceased massive star. It may also have spent some time at the center of a large primordial minor planet that was destroyed by a major impact early in the history of our own solar system. This sixty-ton, nine-feet-wide meteorite has not been moved from its discovery spot. The only noticeable crater at the site is the one dug around it so that tourists can get a better look, but it is perhaps possible that some geological processes may have hidden the original natural crater.

If the Hoba meteorite had arrived at seventeen kilometers per second and had impacted into sand at an angle of ninety degrees, we would expect a crater about one hundred meters in diameter. The energy released would have been about the same as that released by four thousand tons of TNT. Therefore, the lack of a detectable crater suggests that Hoba may have had a rather soft, relatively low-velocity landing sometime within the last eighty thousand years. This could also account for its cuboid shape. A low angle entry may be the reason why such a large lump survived to the surface. It may have been slowed down as it skipped through Earth's atmosphere like a stone skipping across the surface of a lake. Its probable soft landing is likely the reason it is the largest meteorite ever found.

No matter how many craters we can measure on Earth, on the moon, and on some other planets and moons, there will always be too many unknown quantities to create a purely theoretical estimate of impact energies. Some of the things that need to be considered are the angle of the impact, whether it hit water or land, the land composition, and the velocity of the impact. The last one might sound simple, but we would need to know something about the relative orbital velocity of Earth and the asteroid at the time of impact, not just the effects of Earth's gravity.

It is relatively easy to drop a marble into sand and measure the properties of the crater formed, so impacts are something that we can test. We can perform experiments to empirically guide impact energy calculations using scalable observed relations. These findings can then be used to make crater simulators like those written by Ross A. Beyer and H. Jay Melosh while they were at the University of Arizona[12]. To reproduce the Chicxulub event, this online crater simulator needs to use a ten-kilometer-diameter iron asteroid impacting a coastline (watery soil) at a twenty-five-degree angle. This produces a crater 250 kilometers in diameter, which is similar to the crater found at the Yucatán Peninsula.

While a massive crater is an obvious effect of such an impact, it is not the only one[13]. For a coastline impact, massive tsunamis would be triggered, and massive amounts of hot rock and noxious gases would be hurled into the atmosphere. The tsunamis would travel thousands of miles and devastate coastlines within a few hours, the hot debris would fall back down and cause widespread fires for a few weeks, and the finer particulate debris would create a thicker, more toxic atmosphere that would block out the sun for a few years. This would cause the average global temperature to drop. Apart from the physical damage of the crater, tsunamis, and fires, the atmospheric changes would destroy a lot of plant life. Soon after, the herbivores, which enjoyed munching on that plant life, would starve, and then so would the large carnivores that enjoyed snacking on the herbivores.

Not everything was lost after the Chicxulub impact, though. While it is estimated that about three-quarters of all species went extinct in that instance, smaller animals were able to survive the post-impact conditions. So, should another ten-kilometer-diameter iron asteroid impact Earth, it is likely that some life would survive. Regardless, it would be the end of Earth as we know it and is something that we should work hard to avoid. Plus, there will come a point where the size is so big that it won't really matter either way; an impact will cause a completely fatal amount of damage no matter what.

Impacts between solar system bodies are clearly not an uncommon event. In fact, many other impacts have been observed on Jupiter and the moon, and we have seen new craters appear on the surface of Mars[14]. If it weren't for Earth's atmosphere vaporizing most of these objects, there would be far more than the almost two hundred documented impact sites. But the vaporization of these objects can also be destructive. Both the 1908 Tunguska and 2013 Chelyabinsk meteors exploded in the air but still caused widespread ground damage[15,16].

Despite the slow metamorphosis of our planet's surface, we have still managed to find a handful of old and large craters on Earth, such as those at Chicxulub in Mexico, Vredefort in South Africa, Sudbury in Canada, and Lake Acraman in Australia. However, some are far more obvious. This is typically because they are much more recent. In fact, we can use cratering to infer the ages of planets' and moons' surfaces. Classic crater examples on Earth are the fifty-thousand--year-old, 1.2-kilometer-wide Barringer crater in Arizona and the three-hundred-thousand-year-old, eight-hundred-meter-wide Wolfe Creek crater in Western Australia[17,18]. Iron asteroids about fifty meters across probably caused both. While these examples did not have any worldwide implications, one only need transpose their diameters across any city to realize that even a "small" unpredicted fifty-meter-wide object could cause a catastrophic loss of life. It might have an incredibly low possibility of happening, but it is not zero.

By now, the case for finding as many chunks of solar system debris as possible should be clear. Impacts can cause a massive amount of damage, and they put people's lives at risk. We need to know if any of them pose a threat to either all life on Earth, or some of it, and figure out how to reduce or remove the risks. These needed efforts might not seem that urgent right now, but just because an impact will probably not happen *in your lifetime* doesn't mean plans can't be made starting today to help out future generations. Unfortunately, perhaps because of the "it's not my problem" issue, there doesn't seem to be a massive amount of interest from most people. One way that could change is if astronomers could show the world the destructive power of actual impacts in real time.

Comets

We now have direct data on the actual destructive power of real impacts. Many have been seen in real time. The most well-known impact witnessed is that of Comet Shoemaker-Levy 9 into Jupiter[19]. The nuclei of comets are not as tough as asteroids, so as Shoemaker-Levy 9 orbited Jupiter, it was torn into smaller fragments by the giant planet's gravity. What started out as a comet perhaps two kilometers across ended up as a series of twenty-one chunks[20]. The world watched as the chunks slammed into Jupiter over the course of a week in July 1994. And while the main events happened on the far side of Jupiter, the largest chunks left scars in the Jovian atmosphere that had almost the same diameter as Earth. However, these chunks had impact energies significantly less than the estimate for the K-Pg asteroid. Compared to our K-Pg asteroid, the Shoemaker-Levy 9 fragments were a lot smaller, softer, and more numerous, and they impacted a gaseous atmosphere rather than a coastline.

There is a rich history in the study of comets. They mostly present themselves to the naked-eye observer with beautiful tails streaming away from the sun, but in reality, they are dirty balls of ice. They are the leftover debris from the cloud of gas and dust from which our solar system originally formed. As comets get closer to the sun, the increasing heat begins to vaporize their nuclei, and streams of gas and dust glint fabulously in the sunlight to court our superstitions and intellectual curiosity. But because comets have been observed for the whole of humanity, and because the majority of our species held on to childish notions of deistic vengeances, comets were typically feared as the causes of terrible events to come.

Objects that appeared predictably, such as the sun, the moon, the planets, and the stars, did not provoke a sense of fear. These objects became revered, especially when ancient astronomers realized that they could be used to predict the seasons for agricultural purposes and to improve our quality of life. On the other hand, comets appeared suddenly and irregularly. They represented a contradiction to the orderliness of the other celestial bodies. Their unpredictable nature played into the hands of those charlatans wishing to manipulate the doubts and fears of the general population. That is, until astronomers stepped in.

Edmund Halley (1656–1742) used historical records to determine that there was a potential periodicity to some comets and was able to predict the appearance of the comet that now bears his name, Halley's Comet. Unfortunately, Halley died sixteen years before his comet returned to prove him right, but his verified calculations meant that people could begin to calm down about these things. The enigma was beginning to wash away. Halley's Comet will return again in 2061 and, in 2134, is expected to pass within thirteen million kilometers of Earth at a relative velocity of about seventy kilometers per second. In this case, its tail could stretch over twenty degrees across the sky. This will be an extraordinary sight for those around to see it.

Nowadays, comets are much less mysterious, but they still pose a threat. We have successfully sent several probes to these objects, including Halley's Comet. The Soviet *Vega 1* and *2* spacecraft, which launched in 1984, dropped probes on Venus and then went on to Halley's Comet where they both passed within nine thousand kilometers of the nucleus. The *Giotto* spacecraft, launched by the European Space Agency (ESA) in 1985, went even closer and passed within six hundred kilometers of the nucleus. *Giotto*, named after the architect and painter Giotto di

Bondone (1267–1337), who depicted a comet in the place of the Star of Bethlehem in the classic nativity scene, also passed through the tail of Halley's Comet. After being pummeled by debris ejected from the nucleus, *Giotto* managed to determine that the tail is mostly water and carbon compounds. There have been several more recent missions that have managed to visit comets, including NASA's *Deep Impact* probe, which purposefully crashed into Comet 9P/Tempel, and ESA's *Rosetta* mission and *Philae* lander, which landed on Comet 67P/Churyumov–Gerasimenko[21]. While these extraordinary missions have allowed us to learn a great deal about comets, we have yet to ask one of the simplest questions: "Where do comets come from?"

As the inner solar system condensed into the sun and the planets, moons, and asteroids, there was residue refuse left behind in huge distant regions now known as the Kuiper Belt and Oort Cloud (named after Gerard Kuiper, 1905–1973, and Jan Oort, 1900–1992). The Kuiper Belt is the source of short-period comets that can take up to a few hundred years to orbit the sun (like Halley's Comet), and the Oort Cloud is thought to be the source of long-period comets that have orbits of hundreds of thousands of years. Out there, in the cold, dark depths, temperatures are so low that molecules such as carbon dioxide and methane, which are typically known to us as gasses, are frozen. These molecular ices, as well as water ice, bind together small lumps of rock and metal that are the nuclei of comets. A chance encounter with another comet, or a gravitational "poke" from the passing by of a nearby star, will send these nuclei tumbling toward the inner solar system.

Since the Kuiper Belt extends out to around fifty times the average distance between Earth and the sun, it can take hundreds of years for short-period Kuiper-Belt comets to reach the inner solar system. Once they get here, these comets could simply plough into the sun. NASA's Solar and Heliospheric Observatory observes such events regularly[22]. These comets could also impact a planet or have their orbits circularized by the gravitational force from a planet. In this case, rather than heading back to the Kuiper Belt, the comets' paths are modified so that they orbit much closer to the sun. Comet 9P/Tempel and 67P/Churyumov–Gerasimenko are both examples of these types of comets. Comet Shoemaker-Levy 9 was also a member of this family.

The Oort Cloud has yet to be directly observed, but based on the orbits of long-period comets, it likely extends out at least one hundred thousand times farther than the average distance between Earth and the sun. This is halfway to the nearest star. Once long-period comets get to the inner solar system, they too could simply plough headfirst into the sun, impact a planet, or have their orbits changed. It takes hundreds of thousands of years for long-period comets to make their way toward the inner solar system, and by the time they get near Earth, they are traveling very fast. This means we may have to react very quickly if one of these objects needs to be redirected to avoid an impact with Earth.

Water

Water is found in asteroids and comets, and it is an incredibly important resource. It is essential for life, and it can be made into rocket fuel, but from Earth, it is difficult to get into space. That's not to say we necessarily need to get it from Earth to space, because water has been found throughout our solar system. It is on the moon, in and on other moons, on Mercury, on Mars, in and on asteroids and comets, and on dwarf planets like Pluto.

Typically, asteroids contain large amounts of hydrated minerals (rocks with water inside), and comets also contain ice. Both of these objects could have delivered water to Earth during the solar system's teething phase[23]. So, while the medieval soothsayers spouted drivel about astronomical harbingers of death, it turns out that comets and asteroids could have actually been responsible for life being able to get going, and keep going, on Earth. The water they contain could also help us as we send people to explore more and more of our solar system.

As we've come to understand asteroids and comets much better, their role in the future of humanity has begun to change. Yes, they are a threat, but they are also an opportunity. In fact, they may be such a great opportunity that private individuals have established businesses to go after them, possibly sparking a next-generation space race driven by the private sector. The commercial interest in space-based water goes far beyond satisfying the thirst and sanitation needs of astronauts. Not only will it provide the oxygen needed by the human body, but it can also be used to supply the fuel needed by spacecraft to launch, maneuver, and traverse the vast distances between asteroids themselves, planets, and eventually stars. With the water they hold, asteroids and comets may hold the keys to our descendants' future.

Dihydrogen monoxide and H_2O are two equivalent ways of describing water. It contains two parts hydrogen to one part oxygen, and—if it is not needed for drinking, washing, or growing things—we can break it into these two constituents using electrolysis. In this process, two metal probes are placed in water, and an electric current is sent between them. The reactions at each probe pull apart the hydrogen and oxygen atoms. The oxygen itself interests astronauts wishing to breathe, and the hydrogen and oxygen can be processed into rocket fuel.

Rocket fuel can be solid or liquid. Solid rocket fuel is similar to gunpowder. The space shuttle Solid Rocket Boosters (SRBs) used a fuel mixture of ammonium perchlorate, aluminum, and iron oxide. Solid rocket fuel can be difficult to work with, because once it's ignited, there is no way to halt the burning in a controlled fashion. SRBs are essentially huge fireworks. Nevertheless, the amount of thrust they produce is staggering, so they are the preferred method of getting things off the ground and up to about sixty kilometers in altitude.

SRBs offer power, but we require space vehicles with controllable thrust in order to achieve the required orbits and perform the necessary orbital maneuvers. More versatile fuels are therefore a must. The space shuttle's main engines used liquid hydrogen (LH2) and liquid oxygen (LOX). These were combined at variable rates in a combustion chamber and thus allowed thrust regulation. But liquid fuels are not without their own problems. Hydrogen and oxygen need to be kept incredibly cold in order for them to stay in liquid form, and that makes them difficult to work with. However, unlike the solid rocket fuels, both hydrogen and oxygen will be in plentiful supply once we can successfully mine water from asteroids and comets.

Whether using solid, liquid, gas, or some other thrust system, the amount of fuel required for each and every mission must be carefully considered. Not only must there be enough fuel to successfully carry out the mission; the cost of lifting mass into space from the surface of Earth is massively expensive. From a financial point of view, if all the mission fuel is to be supplied from Earth, it quickly becomes prohibitively expensive to send large spacecraft into deep space.

Further complications with calculating the amount of fuel required arise when we realize that the mass of the space vehicle is going to decline as the fuel gets used up. In addition, once a fuel tank is empty, it is simply an anchor that needs to be let go. This is why we see rockets jettison most of their bulk shortly after launch.

Helpfully, there is a *rocket equation* [24] to describe the relationship between the mass of the fuel used and the amount of velocity the vehicle gains. This change in velocity is referred to as the *delta-v*. While the delta-v determines the amount of fuel needed, the calculation of the delta-v needed to get between two points must be done first. As we did for the impacts of asteroids and comets, this is done by considering the energies involved. Once these calculations are done, it may very well turn out that it is impossible to complete a mission with the vehicles we might have. In order for the mission to go forward, we may be forced to process the water found in asteroids, comets, and the moon into liquid rocket fuel.

At the most fundamental level, space travel is all about energy. Sure, we need to try to make our astronauts somewhat comfortable, but if we want to get some mass, *any mass,* to some other point in the solar system (including nudging the orbit of an asteroid), we must start with calculating the energy requirements needed to get that done. To demonstrate this, we can ask a straightforward question: How much energy is required to put one liter of water in a low-Earth orbit? The astronauts in the International Space Station (ISS) need to drink, after all.

The ISS orbits at an altitude of about four hundred kilometers and travels at about 7.7 kilometers per second. On the ground, our liter of water is at an altitude of zero meters and travels at the rotational speed of Earth, which is about 0.5 kilometers per second at the equator. The energy needed to lift one liter (one kilogram) of water from zero to four hundred kilometers is about 4 million joules. The joule is a unit of energy named after James Prescott Joule (1818–1889). The energy needed to get one liter (one kilogram) of water going from 0.5 kilometers per second to 7.7 kilometers per second is about 30 million joules. In total, 34 million joules is equivalent to 8 kilograms of TNT. This is the reason we need huge rockets to get things into orbit and why it is so expensive. The mass of fuel needed to launch something into space is going to far exceed the mass of that something, so the more fuel we can find in situ, the better.

The delta-v required to get into low-Earth orbit is actually greater than the final orbital velocity. To get to the ISS at 7.7 kilometers per second requires about 9.5 kilometers per second of delta-v. This is due to the air friction (drag) as the rocket punches through Earth's atmosphere. All in all, to get one kilogram of water to an ISS astronaut, we need about nine kilograms in total.

These fairly straightforward calculations show that to get into an orbit four hundred kilometers above Earth, about 90 percent of a rocket's total mass needs to be fuel. Even once we get into orbit, fuel is needed to make changes to the orbit. For ISS, this is typically dodging space debris and correcting for the tiny amounts of atmospheric drag. However, by the time a rocket reaches orbit, most of the fuel is gone, along with the jettisoned fuel tanks. Without a ready supply of fuel, we can't go much farther.

If we want to go farther than low-Earth orbit, more delta-v, more fuel, and bigger rockets are required. The Saturn V rockets, for example, managed to transfer to the moon less than 2 percent of the total mass that they had at launch. Once in orbit of Earth, a moon landing will typically require about another six kilometers per second delta-v from low-Earth orbit, but that really depends on how quickly we want to get there. A faster trip will need a larger delta-v, both to get up to speed and to slow down again once we get there. The total delta-v needed to land on Mars is about another eleven kilometers per second.

The delta-v required to get to the moon and Mars is quite high, but there are ways to maximize the use of fuel. Mars has a thin atmosphere that can be used to decelerate a spacecraft without burning fuel. Gravity assists from the other planets, where the planet's orbital motion tugs on the spacecraft, can also be used to accelerate spacecraft, but this will add a significant amount of travel time to any mission. Nevertheless, the delta-v needed to get to the moon and Mars is higher than that needed to land on a large number of asteroids and comets. Of the nearly thirteen thousand asteroids cataloged by NASA that are near Earth, over six hundred have delta-v values (from low-Earth orbit) of less than five kilometers per second [25,26].

The *Rosetta* mission to Comet 67P/Churyumov–Gerasimenko demonstrated that we are able to rendezvous with, and touch down on, a comet. To get there, *Rosetta* took a rather interesting tour of the inner solar system. This tour helped with the delta-v requirement for getting to 67P by using several gravity assists from Earth and one from Mars.

When *Rosetta* made its rendezvous and dispatched the *Philae* lander for a controlled touchdown, we discovered that the surface of 67P was covered with about a foot-thick layer of dust (likely carbon compounds). No substantial water ice was found. Water vapor, however, was detected, so ice is likely locked beneath the surface[27]. 67P has spent a lot of time relatively close to the sun, and the radiation from the sun will have affected its surface. So, despite some media claiming the contrary, the results from this one atypical comet do not mean we can rule out comets as primordial suppliers of water to Earth.

The next technological step is to successfully return to Earth a significant sample from the surface of an asteroid or comet. From this, we will begin to determine what might be necessary to mine water from these objects. There have been previous attempts at such a sample return mission, but the results have been limited. In 2005, the Japanese *Hayabusa* spacecraft visited Asteroid 25143 Itokawa. There were some complications during the final descent, communications were lost for some time, and the sampling probe malfunctioned, but the spacecraft did indeed bump into the surface of Itokawa. *Hayabusa* successfully returned to Earth in 2010 and crashed down in South Australia. Detailed and careful inspection of a protected capsule that was part of the original spacecraft (most of which was destroyed as it burned through Earth's atmosphere) revealed about fifteen hundred grains from the surface of Itokawa. The minerals found match the composition of meteorites. Therefore, the *Hayabusa* findings confirmed that the meteorites we occasionally find on the surface of Earth do indeed come from chunks of asteroids.

A second mission, *Hayabusa 2*, is currently on its way to Asteroid 162173 Ryugu, and it is scheduled to return to Earth with a sample in late 2020[28]. NASA's *OSIRIS-REx* mission is another example[29]. This spacecraft, which quite possibly has the most convoluted acronym of any human space mission to date (Origins, Spectral Interpretation, Resource Identification, Security, Regolith Explorer), is scheduled to rendezvous with Asteroid Bennu in late 2018 and return a sample of its surface material by 2023. If these missions prove successful, we will soon be in a position to take the next technological steps. These will be to try and alter the orbit of an asteroid and to send equipment to test whether we can process the in situ resources into useable water. NASA's proposed Asteroid Robotic Redirect Mission (ARRM) aims to do this[30]. If things go according to plan, perhaps we will start mining asteroids for water by 2030.

The *Hayabusa*, *Rosetta*, and *OSIRIS-REx* missions are still very small-scale compared to the asteroid, comet, and Mars missions that are to come, and when crews become involved, water is not only going to be useful, it will be essential. It is incredibly difficult, and likely unrealistic, for us to bring all the water we need to survive into space. However, we know that there is water already up there, and, at some point, we are going to have to find it and use it. This means there are now two major reasons that motivate searching for near-Earth asteroids: We want to make sure none will hit us, and we want to find those from which it will be easiest to extract and process water.

Finding the Valuable Ones

Regardless of whether asteroids and comets supplied Earth with the water currently held in its oceans, it is clear that they contain significant quantities of rather useful materials. In a future where the demands on Earth-bound resources could outweigh what is producible, asteroids and comets may prove essential.

By rendezvousing and landing on comets and asteroids (things that we've already done), we can do several major things. First, we'll be able to alter their orbits. Should we find one on a collision course with Earth, we could subtly push it a little in order to make sure it misses. Caught early enough, the changes in the orbit needed for it to miss Earth are relatively minor. Alternatively, should we find one with enough interesting materials to make it worth exploiting, we could alter its orbit so it moves into a new, stable orbit around Earth or the moon. This would cut down on the amount of commuting necessary to bring the collected resources back to Earth. Second, whether the object is left on its original orbit or put into orbit around Earth or the moon, we'll still be able to process the materials in situ to produce fuel in space and, perhaps, supply other demands back on Earth. An asteroid or comet could become the first space-based service station and provide water, fuel, and building materials.

Both orbital modification of asteroids and comets, and the mining of materials from them, are achievable goals. However, how do we find them, how do we know we've found them all, how do we calculate their orbits, how do we know if they pose an impact threat, and how do we know what they are made of?

The ones that we are very interested in are the so-called Near-Earth Objects (NEOs). Finding them takes either patience or luck. Asteroids are mainly contained to within a few degrees of the plane of the solar system, much like the planets, but comets could come from any direction. They could also be moving really quickly. This makes it challenging to rendezvous with one and perhaps modify its trajectory enough to somehow make it safe.

Regardless of the challenge, the only way we are going to find them is if we monitor the whole sky for their signatures: faint pinpoints of light moving against the background stars. Somewhat like the planets themselves, NEOs look like faint wandering stars.

The surfaces of both asteroids and comets can be quite dark, so they typically don't reflect much light. This makes them very faint and means that, unless we are using a really big telescope that collects a lot of light, we simply may not spot them all. However, there are NEO search programs funded by NASA that network underutilized small telescopes[31]. These telescopes generally have large fields of view for maximizing the areas of sky that can be monitored, but they still struggle to detect the really faint objects that have diameters below one hundred meters. On top of all that, these telescopes are only used for NEO hunting a fraction of the available time when perhaps they should be entirely dedicated to it.

There is another limit imposed on the discovery of NEOs that has nothing to do with the instrumentation used, and that is how fast the data can be inspected for NEO signatures. A telescope can be programmed to automatically carry out a search pattern without needing to be nursed through the night by a telescope operator. It means that every morning, instead of going to bed, astronomers wake to thousands of images and hundreds of gigabits of data that need to be analyzed. Computer programs whittle down this data to disregard anything that is not interesting, but even then, there can still be hundreds of by-eye inspections to make.

In general, computer programs cannot tell us, "This is a new NEO." They can only tell us, "This *might* be a new NEO." A program may have flagged a piece of space-junk, a satellite, an already-known NEO, some other astronomical event, or just a blip from an electronic gremlin in the instrument itself. While a team of astronomers can make some headway into this mountain of data, an army is what's really needed.

A PhD in astrophysics is not needed to learn what an NEO signature looks like—just a little specific training. The same can be said for classifying what galaxies look like. This means that non-expert citizens can become part of a large science team focused on a specific goal. The success of astronomy-related citizen science is one of the direct pieces of evidence that demonstrates the enthusiasm people have in the subject. The Zooniverse project gathers a whole host of science goals that require the deductive reasoning of many human brains on vast datasets[32]. It hosts the Asteroid Zoo to help hunt for resource-rich asteroids and the Galaxy Zoo to help determine the nature of galaxies. Participants can also help study chimpanzees, cyclones, and cancer.

Currently, there are eleven astronomy-related Zooniverse projects and over one million registered citizen scientists. In the future, these numbers are almost certainly going to increase, especially when facilities like the Large Synoptic Survey Telescope (LSST) begin operations[33]. Construction of this telescope began in 2014 at the Cerro Pachón observatory in Chile. It is 8.4 meters in diameter and has been designed to provide a very large field of view to see as many objects as possible in one go. The mission of LSST is simple: Survey the sky continuously for ten years and let other people figure out what science can be done. This is opposite to the normal approach of having a question that we want to answer and building something new to answer it. It is currently scheduled to begin operations in 2023, and it is predicted to generate fifteen terabytes of data per night. Searching this data for NEOs, or any other astronomical objects, is going to require a monumental effort in which citizen scientists will be able to play a fundamental role.

Once an NEO has been found, the next task is to calculate its orbit. This is unlikely to be done by the survey telescope that spotted the thing. Once people know there is something interesting to be done, several other telescopes will carry out follow-up observations. Its motion through space will be monitored over the course of several weeks, or months, in order to build the full three-dimensional (3D) path on which it is traveling.

The longer an NEO is observed, the more accurate its calculated path will be. With only a handful of quick measurements, the uncertainty in its future positions will be large. Nevertheless, this does not prevent us from running the clock forward to see if this new NEO might at some point in the future collide with Earth. However, even with high-quality data on its current path, the uncertainties in its future positions will get larger the further into the future the calculations are stretched. This means that we can place a probability on each NEO as to whether it poses an impact threat. We can also calculate whether it has a favorable delta-v for a visit and whether we may be able to alter its orbit to prevent disaster or provide commercial mining opportunities.

The impact threat is not just dependent on whether the object will collide with Earth. We really wouldn't care too much if an NEO only one meter across were going to hit us. The likelihood of any damage resulting from such an event is essentially zero, as it will burn up in our atmosphere. At a 1999 NEO conference in Turin, Italy, a collection of astronomers and planetary scientists voted on an easy way of categorizing the hazard that a discovered NEO poses to Earth. It is known as the Torino scale, because that is how Italians pronounce Turin, and it goes from zero (we don't care), to one (that's normal for a brand new discovery), to ten (we're totally Chicxulubed). There is a sliding scale of probability and destruction between the two extremes. Projected out over the next hundred years, all known NEOs have a zero on the Torino scale [34].

Asteroid 99942 Apophis has caused the biggest worry in the past[35]. At one time, it was at level four on the Torino scale, meaning it had about a 1-percent chance of impact and astronomers should keep an eye on it. Despite its threat, this asteroid piques the interest of people designing mission concepts to test asteroid deflection methods, and now that the asteroid mining game has begun, 99942 Apophis could be tapped as an in situ resource. Before that can happen, however, we really should work out whether or not it is worth a visit, and then we need to place some constraints on what it is made of.

NEOs are quite cold, so the signal coming from them is essentially reflected sunlight and a little thermal infrared.

Nevertheless, a lot can still be learned from that reflection. We can look at the reflected spectrum and compare that to the spectrum of the sun. Any differences must be attributed to the surface properties of the asteroid. We can also compare reflectance spectra directly with those from meteorites. When we find a meteorite that reflects light in the same way, we can test what it is made of and infer these materials are present in the observed asteroid. Meteorites are former NEOs, after all.

With the addition of their tails, comets can be studied in similar ways. The tails both scatter sunlight and intrinsically glow. Either scenario offers the chance to figure out which atoms and molecules are being ejected by the nucleus. From what we have seen, comets and asteroids are comprised of both simple and complex molecules, as well as a lot of metals. That makes them ideal one-stop shops for astronauts traveling around the solar system.

We don't need to fly a spacecraft to comets and asteroids to get an idea of their basic parameters like their size and shape. As with the tails of comets intersecting the light from background stars, asteroids and comets themselves can pass in front of background stars and momentarily block out that light, which is called *occultation*. Astronomers regularly apply this approach to determine the nature of planets around other stars, as they can also block out the light from their host star. But applying this method to NEOs is tricky for a number of reasons. If an asteroid or comet passes in front of a star, the NEO's shadow is essentially sweeping across Earth in precisely the same manner as the moon's shadow sweeps across Earth during a total solar eclipse, just on a much, much smaller scale and very, very quickly. It happens so fast that astronomers need a high-speed camera to get the time resolution required to place strong constraints on the object's size. The shapes of NEOs are not regular, either; therefore, any data we get from a single event would only provide enough information to place loose constraints on their sizes and shapes. Quite a few occultations would need to be recorded before we could begin to be sure about their sizes and shapes.

No matter how good we get at detecting NEOs, the smaller the NEOs are, the harder they are to spot and characterize. We aren't really worried about the small ones, though. Nevertheless, there are still a huge number of NEOs to be discovered, so we still have a lot of work to do. Hopefully, one day, we will be able to relax, knowing that any imminent catastrophic collisions can be avoided and that those NEOs with favorable delta-vs for visits, or for being parked into a stable Earth orbit, are being mined for their in situ resources. The extraction of resources from asteroids and comets may well be humanity's next great gold rush.

Chapter 3: Unavoidable Problems

The main reason asteroids and comets are avoidable problems is their size. Compared to any other astronomical body (planets, stars, galaxies, our universe itself) they are absolutely tiny. The difference in scale of a one-kilometer-wide asteroid and the sun is the same as the difference in scale between a small grain of sand and the height of the Empire State Building. The energy requirements needed to manipulate such an asteroid are achievable. The energy requirements needed to manipulate all the other astronomical objects are not. The problems presented to us by our sun as well as by our universe are unavoidable.

The Fate of the Universe

A little clarification is needed when stating that the problems presented by the sun are unavoidable. The evolution of the sun, as we will find out later in this chapter, will render Earth uninhabitable in a few billion years. In terms of the astrophysics, this is an unavoidable problem. But that doesn't mean we have to be on Earth when the sun gets to that point. As we begin to pursue the opportunities presented by asteroids, comets, moons, and other planets, we will begin to migrate through our solar system and beyond. Our sun will eventually scorch the surface of Earth, but if we play our cards right, some of our descendants aren't going to be anywhere near it when it does. We can get away from it. Our universe, however, presents a different story. It too is evolving, and we can't get away from that. Whatever is going to happen to our universe in the future is truly unavoidable.

Our expanding universe is about fourteen billion years old. The expansion is not noticeable on the scales that we experience every day, but it is noticeable when we look out across our universe at objects that are incredibly distant. On very large scales, almost everything is moving away from everything else.

The large-scale structure of our universe is incredibly counterintuitive, the timescales difficult to grasp, and we constantly fall on metaphors in order to try to make sense of things. On paper, we have no problem using mathematics to deal with situations in which more than three dimensions are present—we can simply add another term to the equation. This is something that the human mind struggles to comprehend. Just try imagining a four-dimensional (4D) shape. This is the type of mind bend needed if we want to visualize a complete geometric picture of our universe.

What is our universe expanding into? This common question goes some way to highlight the problems we generally have with understanding dimensions, and it shows how we sometimes struggle to ask meaningful questions. We cannot give a useful response to that question, because the question itself conforms to our everyday experiences, whereas our universe does not. Consequently, the fairest response to the question is, "We don't know."

To estimate the fate of our universe, we don't need to overcomplicate things. Based on our current observations and understanding, we can predict what the consequences of the expansion will be by using our everyday experiences and some basic science to figure it out. This science can appear in both introductory physics and chemistry classes.

Astronomy Saves the World

There are three fundamental properties that a region of space can have: pressure, volume, and temperature. The relationships between these three things are well known. If the volume increases, the pressure decreases; if the pressure decreases, the temperature decreases; and if the temperature decreases, the volume decreases[1]. Anyone that has ever inflated or deflated a tire will have some experience with these relations. The volume of a tire is fairly constant, so when air is pumped in, the pressure goes up. Because it takes a while to pump up a tire, the system isn't entirely isolated. There is time to exchange heat with the surrounding environment, so there are no wild temperature increases. However, when air is quickly let out of a tire, the system doesn't have time to lose much energy, and the air coming out gets really cold. The pressure drops rapidly, and so must the temperature.

We can use these relationships as an analogy to understand the fate of our universe. As far as we know, it is an isolated system, but because of its expansion, the volume is increasing. This means that the average temperature must be decreasing. As the expansion continues, our universe will basically get colder and colder and colder and colder. Eventually (and to understand exactly what that eventually means is tricky), after trillions and trillions and trillions and trillions of years, all stars will have turned off, fundamental particles will have become incredibly rare, and even black holes will have evaporated. Our universe will have such a low temperature that essentially nothing can happen within it. We have no idea if life (as we know it) can exist in these conditions, but if it can, it would likely die of boredom. Perhaps our universe will reach an energy state where conditions are weird enough to generate new universes, and perhaps this is how our universe came to be in the first place, but suffice to say, the fate of our universe is a slow, cold death.

Nuclear Energy

Returning to the present, we find that there are plenty of places in our universe where there is a considerable amount of heat. The sun is a good example. Typical stars exist in a state where the forces involved are essentially balanced. Gravity, as usual, plays a critical role. But if there were nothing to balance gravity, the sun would collapse and form a black hole, enshrouded by an event horizon (a kind of physics no man's land), about four miles across.

An obvious thing about stars, more obvious than their gravity, is that they are bright. This is especially obvious for the sun. There is a lot of light being generated somehow, so the sun must be powered by something. But what is it? We know how far away the sun is, we know how massive it is, and we know how bright it is, i.e., how much energy is coming from it, so it is easy to rule out anything that isn't really, really powerful.

The most powerful artificial explosion to date was on October 30, 1961. It was triggered by the Soviet Union during the test detonation of a thermonuclear weapon on the island of Severny, way inside the Arctic Circle. While this was not the largest explosion ever experienced by planet Earth—that honor goes to one of the many asteroids that have collided with Earth over the eons—it was the most energy purposefully generated in a single event. It had the equivalent explosive force of fifty million tons of TNT. Given the name Tsar Bomba, the device responsible weighed twenty-seven tons and used nuclear fission (atom splitting) to trigger the nuclear fusion (atom sticking) of hydrogen. The power source for this totally destructive device was the same power source that provides the gravity-balancing and life-sustaining energy of the sun. It also gives us cancer. Before we can discuss how nuclear energy ties into how stars work, and how our star is going to render life on Earth impossible in the future, we must first understand a little bit of nuclear physics.

What is the smallest thing you can think of? Halve it. Now what is the smallest thing you can think of? Halve it again. Could we play this game forever, or will there come a time when we find something that cannot be divided in half again? This is actually a very old game that was played by two ancient Greeks, Leucippus and Democritus (c. 460 BC–c. 370 BC). Their take on this was to suppose there was something fundamentally small and indivisible that makes up all larger things. They called these particles atoms.

Hydrogen is the simplest, most fundamental, and smallest of all the atoms or chemical elements. Normally, when not at the center of a star or being zapped by high-energy beams of light, hydrogen floats around wanting to get to a lower energy state. It can do this by giving off light (glowing) or by combining with other atoms. When gravity compresses it and heats it up to the conditions found at the center of a star, hydrogen will undergo the same reactions that took place when the Tsar Bomba was detonated. It will fuse together with other hydrogen atoms and eventually form helium. This is not as straightforward as it may seem.

When cold enough, hydrogen is made up of two particles: a relatively large, positively charged proton and a tiny, negatively charged electron. Although the two particles have different sizes and masses, they have exactly opposite electric charges and attract each other. Together, they make up a neutral (zero total charge) atom. We can think of an electric charge as the little electric zap we get from static electricity. This is the effect of many electrons jumping between our hand and whatever we touch.

There are both positive and negative electric charges just like there are positive and negative magnetic poles (north and south). However, under the core conditions of a star, negatively charged electrons that would normally be bound to positively charged protons have been given a massive amount of energy. Like incredibly excited children, those electrons have become totally unbound and are off partying all over the place. This leaves the hydrogen as just a collection of simple protons.

Helium-4 consists of two protons and two chargeless neutrons (hence the dash 4), which are particles with just a tiny bit more mass than protons. Neutral helium would also include two electrons, but like hydrogen, there is too much energy at the centers of stars for electrons to remain bound. As you can imagine, the likelihood of four hydrogen protons coming together at the right place and the right time to form helium-4 is incredibly small. Instead, there is a multi-channel chain of events that takes place at the centers of stars in order for helium-4 to form.

All these chains start along the same path, because there is essentially only one thing that can happen to begin with: Two protons must collide with enough energy to overcome the natural repulsive force that exists between two positive charges. At the center of a star, there is enough energy for this type of collision to take place, because gravity has crushed the material by a massive amount. Two protons can't just sit squashed against one another indefinitely, because their mutual charges are pushing them apart. Therefore, either the two protons go their separate ways again or one of the protons must use a bit of the available energy to convert itself to a neutral neutron; a small, positively charged positron; and a neutrino.

A positron is the anti-matter equivalent of an electron, the difference being an exact opposite electric charge. It must be positively charged if it is leaving behind a neutron. This decaying process needs a weak force to provoke the positron. That same weak force can also be used to glue electrons or positrons back onto neutrons or protons. Anyway, the emitted positron goes off to join the electron party. The party is so awesome that these positrons and electrons annihilate one another to produce very high-energy gamma ray photons or "particles" of light. That's a pHoton, not a pRoton.

After the positron escapes, the remaining proton and new neutron stay together to form heavy hydrogen (deuterium). Whatever existed in terms of the total charge, mass, and energy before the collision must also exist after the collision, because charge and the total mass-energy cannot be destroyed. Therefore, to fulfill that requirement, a rather difficult-to-detect particle (a neutrino) must be produced.

Once the deuterium is formed, there are still plenty of protons left to collide with it, and now that a neutron is present to glue them together, another proton can happily and quickly join this growing nucleus. This process is also accompanied by the emission of yet another gamma ray.

A two-proton atom is no longer hydrogen. It now has twice the positive charge and over three times the mass of hydrogen. It is a different element called helium. However, the number of neutrons can vary. In this case, there are two protons and one neutron (three total particles), and helium-3 is formed. While the number of protons defines the element, the number of neutrons defines the different isotopes of an element. Helium-3 is an isotope of helium.

The number of neutrons in an atom is very important. They carry a nuclear force that binds the two protons despite their mutual repellence. This nuclear force arises as a result of a strong force that holds together the components of both protons and neutrons (quarks). And this is where things *start* to get complicated.

We've been discussing particle physics for a while now, and there are plenty of people much better equipped to explain this stuff than I am. Regardless, these discussions demonstrate one other interesting fact about astronomy. To understand it, one must have enough knowledge to jump from the absolute smallest scales of our understanding to the absolute largest. To be a modern-day astronomer, one must become a jack-of-all-trades physicist.

Let's return to the helium-3 isotope—two protons being held together by the strong force propagated by a neutron. Another neutron still needs to be introduced to make the more common helium-4, and this is where the hydrogen-to-helium conversion chains diverge.

The most common conversion chain is simply for two helium-3 nuclei to collide. Two helium-3 atoms have two neutrons but four protons. Therefore, during this reaction, two protons and a large amount of energy are released, leaving behind the helium-4 (two protons, two neutrons).

Another chain involves the fusion of helium-3 and helium-4. In this case, beryllium-7, lithium-7, and more helium-4 are produced. First, this process already requires helium-4 to be present. Second, a larger number of total protons and neutrons is needed in order to make these more massive atoms.

A third chain takes beryllium-7 and hydrogen to make boron-8, beryllium-8, and even more helium-4.

Accompanying all these chains is the production of many more gamma rays and neutrinos. Because all these reactions require more and more like charges to be forced together, the amount of energy needed increases. This means the probabilities of these heavier reactions decreases, and the amount of heavier elements produced is low.

The key points from the conversion of hydrogen-to-helium are complicated but can be condensed. The most common process takes six protons to make one helium-4 nucleus. Two protons decay into neutrons and combine with two more protons to form the helium-4 nucleus. The two leftover protons go back into the nuclear soup. Along the way, two positrons, two neutrinos, and a collection of gamma rays are produced.

When the amount of matter and energy that is participating in these processes is considered, these explanations produce predictions that can be tested: The expected abundances of hydrogen, helium, and everything else are lots, some, and not much, respectively, and gamma rays are produced and many neutrinos are spat out. Both predictions have been confirmed by observations [2,3].

Neutrinos posed a problem for quite a while. They rarely interact with anything, so it has been a massive challenge to build instruments capable of detecting them[4]. But once they were first detected, it turned out that only one-third of the expected number registered. As scientists, we love these types of results. Once we convince ourselves that we haven't simply been caught out by a flaw in our methods or instrumentation, it means that we are likely seeing new physics! While being proved right is immensely satisfying, being proved wrong lies at the heart of the scientific method and should generate excitement as well. After the discovery of some new physics, the predicted numbers of neutrinos from the sun have now been confirmed. It turns out that there are three different types of neutrino: the electron-, muon-, and tau-neutrino. As a single neutrino travels from the sun to Earth, it can change from one of these flavors to another, which is a neat little trick that kept many heads scratching when the original instruments could only detect one type.

The gamma rays also pose a slight problem. The sun doesn't emit a lot of gamma rays, although some are produced by solar flares. So where have they all gone? It is important to remember that the energy conditions necessary for thermonuclear fusion are present at the very centers of stars, where the temperature and pressure are extremely high and positively charged protons can be forced together. This makes it hard for a new gamma ray to move around in the cores of stars. There are electrons, positrons, protons, and the occasional helium atom, all zipping around the place getting in the way.

A gamma ray faces the same problem anyone has when trying to leave a particularly crowded party or concert. They can only travel a relatively short distance before their path is randomly knocked in another direction by all the other particles present. They are sent on a *random walk*. The more they walk, the more tired they become, and just like partygoers, the photons start to lose energy. By the time it gets to the surface, that photon is no longer a gamma ray. It has lost a lot of energy and can now be of low enough energy to be visible to the human eye.

Typical Stars

These thermonuclear processes have shown us that the pressure conditions at the center of a star and the pressure that inhibits the free escape of the gamma rays produced are enormous. Therefore, the equilibrium conditions in stars are between their own gravity and the pressure caused by thermonuclear fusion.

This is a tremendous oversimplification, but it is good enough to determine why stars, on the grand scale of things, don't really do that much. For the most part, they are in boring equilibrium. *Hydrostatic equilibrium,* to be precise. This is just a fancy way of saying, "A fluid at rest pushing out balanced by gravity pulling in."

To show why this is an oversimplification, it must be noted that fundamental properties change as we move through the star. As we get closer to the core, the temperature and density go up. To maintain the equilibrium, the pressure must also go up. Moving in the other direction, away from the core, the pressure is going to change enough for rather different conditions to be present in different zones of a star. Consequently, stars must be more complicated than a simple fusing ball of hydrogen. They must have structure. The complexity of the structure depends on the star's mass, which is its most important feature, because that is ultimately what determines their temperature, density, and pressure profiles.

Stars that are less than about half the mass of the sun do not get dense enough to get very hot at their centers. It is still hot enough for the thermonuclear fusion of hydrogen, but it happens much more slowly. Somewhat counterintuitively, low-mass stars live a lot longer, because they are better at conserving their fuel. They sip their hydrogen, don't change very quickly, and don't shine brightly.

Stars between one-half and one-and-one-half times the mass of the sun get dense enough to raise core temperatures to the point where hydrogen can be fused a lot faster. In sun-like stars, more energy is generated in a shorter amount of time. They gulp their fuel, change on moderate timescales, and shine fairly brightly.

In sun-like stars, as the distance from the core increases, the temperature and density decreases until the photons generated in the core can move more freely. They get to the edge of the concert crowd. This region is known as a *radiative zone*. When the temperature drops even more, the solar material begins to behave in a more "traditional" manner, and instead of fusing hydrogen, it basically boils it. This produces the same type of process seen in boiling water, i.e., convection (hot stuff rises, cold stuff sinks). This region has been sensibly named the *convection zone*.

When we look at high-resolution images of the sun at optical wavelengths, we see the tops of these convective cells as a network of bright blotches (the hotter stuff rising) surrounded by darker outlines (the cooler stuff sinking). These are referred to as granules, and they are about one thousand kilometers across[5]. Considering the 1.4-million-kilometer diameter of the sun, there are millions of convective cells making up its outermost layer.

Massive stars are even hotter in their cores, and their fuel must be burned a lot faster to prevent gravitational collapse. They guzzle their fuel, run out quickly, and shine very brightly.

In massive stars, there is even more energy available for isotopes heavier than those produced in the lower-mass stars (beryllium-8, lithium-7, boron-8) to form. Because the like-charge repulsion increases when there are larger numbers of protons present, additional fusion reactions require the hotter conditions that the gravity of massive stars provides.

By fusing helium-4 and beryllium-8, carbon-12 can be created. From here, the ongoing fusion of hydrogen with carbon-12, as well as the resulting elements, makes nitrogen-13, carbon-13, nitrogen-14, oxygen-15, and nitrogen-15. This creates an ongoing carbon, nitrogen, oxygen (CNO) cycle. This also contributes to the increasing number of helium-4 atoms. Massive stars are also where carbon-12 itself can fuse together to form oxygen-16, neon-20, sodium-23, magnesium-23, and magnesium-24. They are also where oxygen-16 itself can fuse together to form magnesium-24, silicon-28, phosphorus-31, sulfur-31, and sulfer-32. Finally, helium-4 can fuse with all these previous elements to form argon-36, calcium-40, titanium-44, chromium-48, iron-52, nickel-56, and zinc-60.

Many of these isotopes are unstable enough to decay quickly into other elements, i.e., they find a way to get to a lower energy state. The twenty-eight protons that define nickel-56 are held together by twenty-eight neutrons. However, there is a lower energy state that a mixture of fifty-six protons and neutrons can have. In these cases, the relative number of neutrons is increased. Therefore, to find these lower-energy equilibrium conditions, nickel-56 (twenty-eight protons, twenty-eight neutrons) converts a proton into a neutron by giving off a positron and electron-neutrino to leave behind cobalt-56 (twenty-seven protons, twenty-nine neutrons). Cobalt-56 then quickly does the same to leave behind iron-56 (twenty-six protons, thirty neutrons). At this point, the processing of heavy elements has reached a peak, because the thirty neutrons in iron-56 really bind together those twenty-six protons without needing to find a lower energy state (iron-56 is "stable"). As a consequence, the production of even heavier elements is going to have to wait for the equilibrium condition in these massive stars to end and a lot more energy to be present.

For me, it is the understanding of the thermonuclear fusion processes in these rare, distant, and massive stars that has produced one of the most profound realizations about our own existence. More so than realizing Earth is not flat. More so than realizing Earth isn't the center of our universe. With the exception of hydrogen and helium, everything we see around us—the devices we read; our own bodies; the things we stand, sit, or lie on; what we eat; *everything*—is made from elements fused together in the cores of long-gone, massive stars that lived fast and died young, each going out in its own blaze of glory. This is not conjecture, a guess, or a faith-based belief. It is fact, real, the actual truth, and it is backed up by a wealth of solid, reproducible, and elegant evidence.

Pulsating Stars

With more detailed observations, we are able to determine several things about the stars. If we know their distances and how bright they appear, we can work out how bright they really are. A faint thing that is close could look like a bright thing that is far away, and it is important to know the difference. We can also see how their brightness changes at different wavelengths, i.e., if they are blue, red, or somewhere in between. This tells us the surface temperature of the star. Entirely opposite to the color-coding on water taps, blue stars are hotter, and red stars are cooler.

When the relationship between the temperature and brightness of stars is plotted onto a graph, we find that most stars fall onto a well-defined, slightly wiggly line known as the *main sequence*. We now know that the main sequence corresponds to stars that are fusing hydrogen in their cores.

However, there are a number of stars that do not fall on the main sequence. Some of these stars periodically, over the course of a few days, increase and decrease their brightness due to internal physical processes. They are pulsating, almost like they are breathing in and out.

In reality, pulsations are the result of a complex situation that involves a large number of variables, and the behavior of photons in stellar interiors is fairly counterintuitive. But, we may start by recalling that the conditions within a star determine how difficult it is for a photon to travel through it (the random walk problem).

Until a photon can escape, its energy remains inside the core of the star. When a lot of energy is being produced, and once those photons escape from the core, they can "push" on the outer layers of a star and make it get slightly bigger. If the volume of something changes, it also changes the density and temperature, which affects how easily the photons can travel through that material. Consequently, more photons get the chance to escape during the period of expansion. Once those photons escape, the support for the expansion—the energy of those escaping photons—disappears. The star contracts again, heats up, produces more photons, and the process starts all over. All the while, the star gets brighter and dimmer in a regular cycle.

The expansion and contraction of pulsating, variable stars have been confirmed by other observations that measure the rising and falling speeds of the stellar surfaces as they cycle through the pulsation processes[6]. Averaged over many cycles, these stars maintain an overall balance, and we can describe them using a time-averaged brightness. That is, until the fuel runs out.

Death of Stars

Regardless of whether a star is a pulsating variable or on the main sequence, sooner or later, it will not have enough hydrogen left to maintain its first equilibrium state. Main-sequence stars will leave the main sequence and go through a rapid succession of briefer and briefer thermonuclear equilibrium phases.

Low-mass stars burn their hydrogen so slowly that they will leave the main sequence last. As their main-sequence lifetimes are hundreds of billions of years, and because our universe is only fourteen billion years old, all low-mass stars are still on the main sequence and will be well into the deep future.

Massive, short-lived stars, on the other hand, are the first to leave the main sequence. They burn through their hydrogen like a supercar on full throttle burns through its fuel. Similarly, like many mishandled supercars, massive stars are going to distribute their components in a rather chaotic fashion all over the surrounding space soon after their engines start. It's lucky for us that they do (the massive stars, not the cars). Without this explosion, later generations of stars would not be born in the gas and dust clouds enriched by their thermonuclear ash. It is this ash that makes planets like Earth, and life itself, possible.

In high-mass stars, heavier elements like carbon, oxygen, silicon, and iron have been processed from hydrogen and helium in the much higher-temperature thermonuclear furnaces. Those elements sink toward the center of the star, with the densest at the bottom surrounded by less and less dense elements as we move away from the center. The same effect is seen when pouring vegetable oil and syrup into a glass of water. The syrup is the iron, which sinks to the bottom; the water is the oxygen, which sits in the middle; and the oil is the carbon, which sits on top. One could even consider the air in the remainder of the glass a representation of the helium and hydrogen. If the oil-water-syrup analogy doesn't sound palatable, just think of a cunningly crafted cocktail instead. A B-52 or a Tequila Sunrise, for a more astronomy-related drink.

After a few tens of millions of years, the hydrogen in the core of a massive star is exhausted, and the initial gravitational support mechanism is lost. The star contracts and leaves the main sequence. This collapse drives the density and temperature of the star up, until helium can fuse and produce a second equilibrium state. There is less helium than hydrogen, and the fusion is more energetic, so this secondary support mechanism only lasts a million years or so. Then the collapse begins again until carbon begins to fuse, then oxygen, then silicon. Each new support element is used faster and faster, a few days in the case of silicon, until a massive amount of iron fills the core. At this point, there is nothing that thermonuclear fusion can do to prevent the catastrophic collapse of the star.

As a result of their massive size, the inner parts of an exhausted massive star collapse toward the wickedly heavy core before the outer parts of the star can react. The speed at which this inner collapse occurs is phenomenal, with the parts closest to the core falling at tens of thousands of kilometers per second. The inner collapsing parts therefore reach the core first and bounce off it. At this point, the core is being gravitationally squeezed more tightly together than ever before. The bounced inner parts, now on their way back out, collide with the remaining outer parts of the star, and the shock of these two materials meeting one another at these speeds sends things into overdrive. Temperatures skyrocket, causing photons to jet around, and this breaks many of the heavy elements back down into protons and neutrons. However, the reverse also happens, and some of the energy produced goes into forming even heavier elements like krypton-86, silver-107, iodine-127, lead-208, and uranium-235. All the while, huge numbers of neutrinos are produced along with monumental amounts of energy. The star has become a new super star—a supernova.

The predictions of the supernova scenario include the production of heavy elements, a huge number of neutrinos, a large amount of energy, and either a remnant neutron star or black hole in the place of the once iron-rich stellar core. These have all been confirmed observationally—there is a whole table of known elements, thousands of neutron stars have been observed in our galaxy, we see some of the effects of black holes, and supernovae are detected regularly in other galaxies at a rate of about one supernova per galaxy every one hundred years. Consequently, it is now accepted that at the end of their fusion-fueled lives, massive stars are destined to become celestial fireworks that follow this core-collapse supernova model [7].

Death of Dead Stars

Core-collapse supernovae are not the only way to get a catastrophic and terminal outburst from a star. Supernovae can also be triggered in an appropriately configured double-star system (a binary) in which one main-sequence star has already died.

In much the same way that each individual human is unique, no two stars are ever precisely alike. This means that one star in the binary will remain on the main sequence, fusing hydrogen in its core, longer than the other. It is not uncommon for binary stars to contain one component still on the main sequence and one that has already belched away its outer layers and died down to become a *white dwarf*—the glowing compact core of a former star. Sirius, a star system as well as the brightest star in our night sky, is a good example of a white dwarf–main sequence binary. In this star–dead star system, the orbits of the two are elliptical, allowing the distance between them to vary between eight and thirty-two times the average distance between Earth and the sun. However, in some white dwarf–main sequence binaries, the two stars can be a lot closer together.

When the separations get to around fifty or so times the diameter of the main-sequence star, the gravitational interaction between the two means that material is dragged off the main-sequence star and into a disk around the white dwarf. As the material in this disk sloshes around and spirals onto the white dwarf, we can observe some cataclysmic, but not catastrophic, variability in the brightness of the system. The ultimate result is that the mass of the white dwarf increases.

White dwarfs are fundamentally different from main-sequence stars. Instead of generating pressure with the thermonuclear fusion of hydrogen, they generate it by tightly squeezing electrons. Regardless, even electrons have their limits. As the white dwarf snacks on its main-sequence partner, the mass of the white dwarf increases, and another tremendously powerful event draws closer.

Unlike main-sequence stars, as the mass of a white dwarf increases, its size actually decreases—gravity crushes it even harder. As the compression on the electrons increases, the speed at which they are pinging around within the white dwarf also increases. When that velocity reaches the speed of light, the equilibrium condition maintained by the electrons must break down, and a massive explosion, called a type Ia supernova, takes place[8]. Subrahmanyan Chandrasekhar (1910–1995) calculated that this happens at a very specific white dwarf mass. This mass limit is now known as the Chandrasekhar limit, and it is 1.4 times the mass of the sun. To highlight how important Chandrasekhar's white-dwarf calculations are, the Chandra Space-based X-ray Observatory was named in his honor in 1998.

Fate of Our Sun

The main-sequence progenitors of white dwarfs are sun-like stars. When the hydrogen runs out in the cores of these stars, the mechanism supporting them from gravitational collapse also shuts down. Gravity begins to win again, and the stellar cores contract, making the conditions more extreme. The densities and temperatures in the cores increase, and the outer parts of the stars puff up to become giants.

Although the cores are now much hotter, the expansion means that the energy coming from the stars is spread out over a larger surface. Externally, the stars get cooler and start to become redder. As with massive stars, the heaviest elements (helium, carbon) have sunk to the centers. These cores continue to shrink until they approach the point at which conditions get extreme enough for helium fusion. This provides the internal energy needed to establish a second equilibrium condition. Before that, however, fusion of the remaining hydrogen that isn't in the core provides a little additional energy to push on the outer layers a bit more. They expand again, cool more, and become even bigger and redder. They are now red giants.

Very soon after the sun-like star turns into a red giant, the core conditions allow the fusion of helium to occur, and the star is once again supported against the inevitability of gravity. Helium fuses at a much higher temperature than hydrogen, and so this second equilibrium epoch is much shorter than the main-sequence lifetime. When the core helium is exhausted, and the core is saturated with carbon ash, the star loses equilibrium again, and the core heats even more. When the remaining non-core helium starts to fuse, this last kick of energy blows away a lot more of the outer regions of the stars, and the effect is stunningly beautiful. The resulting objects are spectacular *planetary nebulae* that have nothing to do with planets. They have fabulous rings of red, blue, purple, and green, and sometimes the remnant stellar core—the white dwarf—can still be seen as a point of light at the very center.

This is the fate of our sun, and it will spell the end of Earth as we know it; our planet will no longer be habitable. On paper, there are ways we could try to modify the orbit of Earth by manipulating asteroids. We could use the gravitational tug of many asteroids to slowly move Earth relative to our sun as its temperature evolves, thus keeping the environment just right for life to carry on. But considering the alternative—simply leaving our solar system—that effort can be spent in more practical ways. As the inevitable expansion of our sun is billions of years in the future, we could argue that there is plenty of time to execute such an ambitious orbital modification plan. Who would be around to carry it out, though? It certainly wouldn't be humans as we are today. We will be long gone. Evolution is slow, but not *that* slow. Should someone want to go to such a length to save Earth, it won't be for any practical reason, as our descendants should be living among the other stars by then. The only value would be for the sentimentality of keeping one's place of cosmic origin preserved.

Astronomy Saves the World

Part Two

The Principles

Astronomy Saves the World

Chapter 4: Astronomers Are Really Good at Staring

Astrophysics is founded on basic principles of mathematics, geometry, statistics, physics, and critical thinking. We have seen that these things can easily get complicated. Many topics cause cognitive trouble because they are counterintuitive, catch people off guard, and contradict popular beliefs.

While the forces, energy, and momentum we have already discussed are a core component of physics classes and apply to the actions and skills of everyone every day, they are not where astronomy really begins. Furthermore, a basic understanding of the avoidable problems of asteroids and comets, and the unavoidable problems such as the evolution of stars, is not where astronomy ends.

The starting points of astronomy and astrophysics are dictated by the data we are able to gather, and that is limited by the age and scale of our universe. Everything is far away, and the vast majority of astronomical objects are old and change slowly. With the exception of things within our own solar system, we cannot visit these things to get a different view, a close-up picture, or sample. All we can do is stare. This is not something most people would care to openly admit, because it does seem a bit creepy, but it is quite literally all we can do. Astronomers are even really good at studying the sun, but it's generally a really bad idea to stare at that.

As we study our universe, we use specialized telescopes, filters, and cameras. With instruments like these, and a lot of other highly specialized equipment, we are able to detect the properties of light itself; cosmic rays, which are actually mostly particles; neutrinos, which are also particles; and gravitational waves, which are basically the vibrations of our universe itself. We also have ongoing experiments that are trying to directly detect, rather than see the gravitational effects of, dark matter[1].

Due to the difficulties associated with detecting cosmic rays, neutrinos, and gravitational waves, the overwhelming majority of astronomy is done by looking at properties of light. For this reason, astronomy is a relatively simple science to get involved with, because anyone can see some of the properties of light with their own eyes. Perhaps this is why astronomy is often thought of as a STEM "gateway" science. The basics of astronomy can be explained to almost anyone with introductory mathematical understanding, and in many cases, it sparks an interest in STEM. The beautiful images from observatories like the *Hubble Space Telescope* mean that astronomy can influence more than just the standard STEM disciplines. It often allows science and art to complement each other, making it truly a STEAM "gateway" (where the "A" is for art). The wonders of our universe also influence music, from Beethoven's "Moonlight Sonata" and Holst's "The Planets" to Muse's "Supermassive Black Hole" and Rainbow's "Stargazer."

There is a huge amount of data that can be gathered by simply staring at an object and determining the properties of the light coming from it. Our deductive reasoning skills have allowed us to take this data and piece together the history, and likely future, of our universe. Instruments such as imagers, spectrographs, and polarimeters supply us necessary data and provide absolutely bewildering answers for seemingly superficial questions.

Astronomy Saves the World

Some Simple Questions

What? Why? How? These are simple questions asked too rarely. Admitting our own ignorance is not a sign of weakness. Socrates would argue it is a sign of wisdom. In many cases, these questions are best asked of yourself. They show inquisitiveness and provoke exploration, experimentation, learning, and enlightenment. These are questions that can be asked publically. They can also be used to expose the ignorance or deception of others. They can be a demand for an explanation.

These are not questions that must be saved for the sciences and scientists, but they are questions that the study of science will provoke time after time. They should be used without embarrassment and to hold accountable those elected officials that make decisions about our quality of life on our behalf. What evidence do we have to demonstrate that giving unlimited access to all firearms will result in fewer deaths from gun violence? Why do we limit the rights of some people and provide more rights for others? How can people without the relevant qualifications make informed and reasonable decisions on developing renewable energy sources?

Astronomy is a great motivator for prompting the what, why, and how questions. What is the moon? Why does the sun rise? How do stars shine? But the simplest questions are where we will start. What does an astronomical object really look like? How big is it? How far away is it? These questions are not necessarily unconnected, and they must be asked if we are to begin establishing our place in this universe.

What an object looks like depends on how big and how far away it is. With the exception of the sun, the planets (including Earth), moons, asteroids, and some comets within our own solar system, we cannot see what objects look like in three dimensions. Since most objects are so far away, it will be a very long time before these objects can be seen from even a slightly different angle.

Generations in the far future won't necessarily have to visit or send probes to these places to see them from a slightly different point of view. This is because Earth is orbiting the sun, and the sun is moving through our home galaxy, the Milky Way. The Milky Way has a central bulge of stars, which are surrounded by a flattened disk of gas, dust, and stars that have an apparent spiral pattern. These properties make it, and other spiral galaxies, spectacularly beautiful. The sun is one star in the Milky Way's disk, and it orbits around the center of the Milky Way. The sun also bobs up and down and in and out of the disk. It is rather like the motion of the ride-on model horses on a merry-go-round. The sun is currently slightly above and moving away from the plane of our galaxy's disk. Eventually, as the sun and Earth get farther from the plane of the disk, there will be less gas and dust to look through. When that happens, the view of the center of our galaxy, the disk itself, and the view of our own universe will become a lot clearer. It will take tens of millions of years for the sun to get that far out of the disk, so who knows who will be around to see it. Nevertheless, future beings on Earth will have the opportunity to get a slightly different view of some of the astronomical objects currently around us without having to build fantastic spacecraft.

Regardless of our future spatial position in our galaxy or our future technology, the most basic property of an astronomical object—its actual shape—needs to be treated very carefully. While the problem applies to all objects we look at, even everyday ones, there are three astronomical examples that readily demonstrate this issue: Earth, constellations, and galaxies.

We know Earth is a member of a large population of planets that are potentially habitable, so we can study it as such. The details of Earth provide us with a lot of guidance on the expected characteristics of other Earth-like planets, such as the presence of protective magnetic fields and surface conditions favorable for life.

These are all quite complex details, but when we consider one of the most basic properties of Earth—its shape—the uninformed, casual, careless, or purposefully ignorant observer could think it is flat. As will be discussed shortly, this is obviously not the case.

Constellations are easily recognizable two-dimensional (2D) patterns of stars on the night sky, but their actual shapes are not that obvious. Orion's Belt, which is a well-known section of the larger constellation, is made up of three easy-to-spot stars, Alnitak, Alnilam, and Mintaka. These stars *appear* to form a straight line and *appear* to have similar brightness. In actuality, Alnilam is nearly twice as far away as, and two times brighter than, the other two. The three stars only appear to be in a straight line and only appear to be all at the same distance. If we were to travel a huge distance through space and look at the three stars from a different angle, we would see a V-shaped pattern. The same effect can be seen when looking at a Frisbee from two different directions. One way, it will look like a disk. Another way, it will look like a straight line.

Galaxies are collections of billions of stars bound together by their mutual gravity and dark matter. Their apparent shapes can also be deceiving. We know that galaxies come in a variety of shapes and sizes. Some appear almost round, some appear shaped like a watermelon, and some have spiral arms and bright central bulges like our own galaxy. In these cases, there are some intrinsic differences between galaxies without spiral arms (ellipticals) and those with spiral arms (spirals). But, if we were to look down the long axis of an elliptical galaxy, which is actually shaped like a watermelon, it would look round. Astronomers classify galaxies based mainly on their 2D appearance. They are placed onto the "Hubble tuning fork diagram" where two classes of spiral galaxy form the two prongs of the fork and elliptical galaxies form the handle[2]. Two galaxies that are exactly the same would be given different Hubble classifications depending on their orientation to our line of sight (end on or side on).

Some Mathematics

The distance to an object plays a huge role in an object's appearance, but when we look at the sky, we do not see any information about that. All we see is where it would be if it were projected onto the inside of an imaginary celestial sphere[3]. This means that astronomy really starts with the mathematics needed to understand spheres and circles.

The circle is the simplest shape, because it is entirely defined by one parameter: its diameter. This particular shape also has an infinite number of symmetries, and as the only 2D object with these properties, many people consider it a perfect shape.

When we divide the distance all the way around the edge of the circle by the diameter, we end up with a number we call *pi* (symbol π, pronounced "pie"). Pi is a never-ending number (3.14159265358979 . . .), so at the core of a shape that is held by many to be perfect lays a number that no one will ever perfectly know. The diameter could start at either end, so the dimension of the circle we typically quote is the *radius*. This starts at one place—the center of the circle—and is half the length of the diameter.

If we stretch out one of our arms, point, and turn all the way around to trace a circle, the distance the end of our pointed finger travels (the circumference) is equal to two times the length of our arm multiplied by π. This is a result of how we defined π in the first place. This also means that if we turn only halfway, the end of our finger has only traveled the length of our arm times π, and so on. There is a relationship between the angle we turn and the length of the arc traced by our finger. We can define the angle as the arc length divided by the length of our arm. Typically, we would use the symbol theta (θ) for the size of an angle, so we can say $\theta = s/r$, where s is the arc length and r is the length of the radius.

The distance to an object can now be treated simply as the radius of a circle with Earth at the center. Therefore, using the

$\theta = s/r$ relationship, and assuming we can accurately determine the distance to an object (r) and can measure the apparent angle on the sky the object subtends (θ), we will know how big that object really is (s). If we already know how big the object is, we can calculate its distance from Earth based on its apparent angle.

It's common knowledge that there are 360 degrees in a circle. In this case, the term "degree" is simply used to let us know that we have been told about an angle rather than some other random ratio. However, the more natural way to define an angle is to base it off the one defining characteristic of a circle, its radius. If we take the ratio of an arc length to the radius, we get an angle in units of *radians*. It follows that there are then 2 times π radians in a complete circle of 360 degrees; knowing this allows us to convert between radians and degrees (one radian equals 180 divided by π, or about 57 degrees).

Understanding the simple circle is the first step in establishing how we define a method for describing not only the apparent size of an object but also its position on the night sky. But there are some more details that take the usefulness of the simple circle even further. This usefulness is applicable not only to astronomy but to mathematics in general, most areas of science, and a large proportion of engineering.

Consider the behavior of the shadow cast by a falling stick against a wall (see Figure 4.1). The shadow is coming from a strong light source placed some distance away from the stick. As the angle between the falling stick and the wall increases, the length of the shadow decreases. When the stick is upright against the wall, the shadow has the same length as the stick. When the stick has fallen to the ground, the angle between it and the wall is ninety degrees ($\pi/2$ radians) and the shadow length is zero. The shadow length does not change at the same rate at which the angle changes. If the stick is falling at a constant rate (the increase in the angle is constant), at first, the change in shadow length is slow. As the stick gets closer to the ground, the shadow gets smaller faster.

Figure 4.1: A falling stick casts a smaller and smaller shadow on a wall as it falls closer and closer to the ground. The relationship between the increasing angle of the stick and the length of the shadow gives us the basics of trigonometry.

With the behavior of the falling stick's shadow in mind, let's now consider the unit circle. This is simply a circle with the radius set to one. Imagine a unit circle as a clock, with the second hand being one inch long. At noon, the second hand is completely vertical and lying along a line we shall call the y-axis. A quarter turn later (fifteen seconds, $\pi/2$ radians, or ninety degrees), the second hand is completely horizontal and lying along a line we shall call the x-axis.

Now, consider the shadow that would be cast by the second hand on the y-axis if a light source were placed far away on the x-axis. At noon, the shadow cast by the second hand on the y-axis is equal to the length of the second hand and equal to the radius of our unit circle, i.e., it is equal to one inch. As the second hand moves toward the three o'clock position, the shadow on the y-axis gets smaller in a predictable way. Close to the noon position, the length of the shadow changes slowly as the second-hand angle increases. Close to the three o'clock position, the length of the shadow changes quickly. The shadow starts at a length of one and goes toward zero. The shadow length goes toward minus one as it moves past three o'clock to six o'clock, back to zero at nine o'clock, and back to one again at noon. This curved relationship between the length of the shadow on the y-axis and the angle that the second hand has moved through is called the *cosine* function.

If a light were placed far away on the y-axis, far above the second hand, the same exercise could be done with the shadow on the x-axis. This time, the shadow would be falling on the floor rather than the wall. When the second hand is at noon, the length of the shadow on the x-axis is zero. The shadow goes to one by three o'clock, zero again at six o'clock, minus one at nine o'clock, and zero again at noon. This curved relationship, between the length of the shadow on the x-axis and the angle that the second hand has moved through, is called the *sine* function.

Roll in all the different ways to describe the angle of the second hand (o'clock, radians, degrees, fractions of a circle), and we find ourselves engulfed by large amounts of information. This is when it is sometimes a good idea to include a picture of what is going on. Instead of having to build a visualization of the situation in your mind, we can see it in print (see Figure 4.2).

Seeing the data in graph form demonstrates some things that are not necessarily obvious from the text. When the second hand has hardly moved from noon, the cosine function is always close to one. For angles close to three o'clock, the sine function is always close to one. If the cosine function were moved forward by ninety degrees (quarter of a revolution or $\pi/2$ radians), it would become the sine function.

In addition to relating the radius and the arc length to the angle rotated through, the unit circle allows us to easily relate the relative lengths of the sides of triangles to the angles within those triangles. This is the basis of trigonometry, and it has been known for thousands of years. It goes back at least as far as the ancient Greek astronomer Hipparchus of Nicaea (ca. 190–120 BC) who, among other things, calculated trigonometric conversion tables for easy referencing.

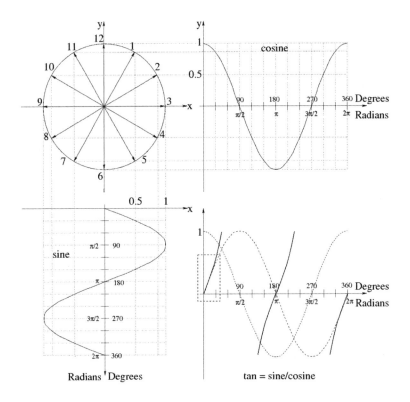

Figure 4.2: The relationship between the axis lengths and the angles that form the basis of trigonometry. The cosine curve corresponds to the falling stick's shadow length on the wall (*y*-axis). The sine function would be the same for the shadow on the floor (*x*-axis).

We can build new functions by combining sine and cosine functions as well. When sine is divided by cosine, as it is in the bottom right-hand panel of Figure 4.2, we create the *tangent* function. Here, the tangent of zero is zero, and the tangent of ninety degrees (quarter of a revolution or $\pi/2$ radians) is infinity (1 divided by 0). Collectively, the sine, cosine, and tangent functions are known as the trigonometric functions. These must become second nature to anyone wanting to pursue a STEM career. The details of the complete 360-degree unit circle, while important, can be reduced to just one quadrant, as in Figure 4.3.

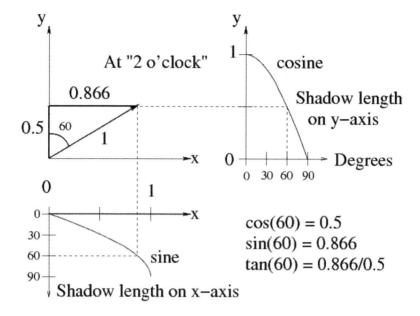

Figure 4.3: One quadrant of the unit circle showing the relationships between the three trigonometric functions. These fundamentals lie at the heart of the STEM disciplines.

We can also apply known geometric relationships to the trigonometric functions that allow us to transform seemingly complex functions into less messy ones. For example, the ancient Greek philosopher and mathematician Pythagoras (ca. 570–495 BC) found that if we take the length of the two shorter sides of a right-angled triangle, square them, and then add them together, the result is the squared length of the long side[4]. In the case of a circle, the long side of a triangle, the hypotenuse, is simply the radius, or $x^2 + y^2 = R^2$, where the radius $R = 1$ for a unit circle. As the x-axis in a unit circle is used to build the sine function and the y-axis is used to build the cosine function, we can insert them here to show that $sin^2 \theta + cos^2 \theta = 1$. There are hundreds more of these trigonometric identities.

What do all these angles have to do with astronomy and how to find the sizes of astronomical objects? How big is the moon? How big is the sun? How big is the nearest galaxy? When we measure their apparent sizes, we can now put the data in terms of angles: degrees, arcminutes (1/60th of a degree), and arcseconds (1/60th of an arcminute or 1/3600th of a degree). The Andromeda galaxy (our nearest big galactic neighbor) extends over three degrees. The Large and Small Magellanic Clouds have apparent angular diameters of about ten degrees and five degrees, respectively. The really bright part of the iconic Orion Nebula is about one degree across. Both the moon and the sun are about half a degree across. The majority of other objects have apparent diameters that are much smaller than this. The angular diameter of Jupiter, for example, is between thirty and fifty arcseconds depending on what time of year we observe it from Earth.

When we measure the apparent angular diameter of an object, we simply build some triangles. The long lengths of the triangle are the distance to the object, and the short length is the diameter of the object. The angle we measure is the angle between the two long lengths of that triangle. In Figure 4.2, there is a dashed rectangle on the part showing the tangent function. Inside that rectangle, both the sine and tangent functions look as if they are almost straight. If we move through a tiny angle like 0.017 radians (one degree), we increase the value of the function by the same amount, i.e., the sine of 0.017 radians is 0.017, and the tangent of 0.017 radians is 0.017. The actual difference between the sines and tangents of small angles when those angles are 0.017 radians is only 0.005 percent. It is rare in astronomy for us to know anything with that type of precision, so for the sake of simplicity, this *small angle approximation* is commonly used.

The power of this approximation is demonstrated when we notice that the angle we measure for an object (θ) is equal to the actual size of that object (d) divided by the actual distance to that object (D), i.e., $\theta = d/D$. There is a reason this may sound familiar.

We saw the relationship when defining the angle in the first place: $\theta = s/r$, where s is the arc length and r is the radius of a circle. Since we know the average distances to the moon (380 thousand kilometers), the sun (150 million kilometers[5]), and Andromeda (twenty-four billion billion kilometers[6]), and their apparent diameters in terms of angles (0.5 degrees = 0.0087 radians, 3 degrees = 0.052 radians), we can easily calculate their actual diameters to be about thirty-five hundred kilometers (moon), 1.4 million kilometers (sun), and two billion billion kilometers (Andromeda). The reason this calculation is so simple is that on a large circle and for a very small angle, the curve in the arc length is very tiny. So tiny that a straight line can easily approximate it. This happens when we stand on the surface of Earth to see what we can see. It *looks* flat when it is in fact curved. This is one fact of geometry that seemingly confuses flat-Earthers[7].

Earth Is Not Flat

When we consider the diameters of iconic astronomical objects, it is easy to forget that we are standing on one. All of us are ready to mock people who claim Earth is flat, but how many of us could actually demonstrate that it is not? Of course, we all know that Earth is somewhat spherical, because we can happily refer to the jaw-droppingly beautiful (and not faked) pictures of a spheroidal Earth from space, can point out that man-made objects orbit it, and can travel all the way around it in an apparent straight line and return to the starting point. But what everyday evidence is there that we can show the average person to prove this fact?

We can watch boats disappear over the horizon as they travel farther and farther away from shore. If Earth were flat, a ship traveling away from us would simply get smaller and smaller. Instead, the hull disappears first and then the mast as the boat moves behind the curvature of Earth. We also know that if we climb to the top of a tall object, we are able to see farther.

In modern times, we can watch a large rocket launch. When these things rise into orbit, they don't go straight up in the air. They follow graceful parabolic arcs that typically disappear toward the eastern horizon. Of course, if Earth were flat, a rocket wouldn't go into orbit anyway.

With today's technology, it is relatively easy to get either yourself, or some recording device, to a high enough altitude to actually *see* the curvature of Earth. There are plenty of examples of people attaching high-definition cameras to weather balloons and video recording what is seen as the balloons rise—a curved Earth. In a large passenger aircraft at a typical altitude of around ten kilometers, the apparent curvature of Earth is about three degrees and noticeable to most observers.

If we don't live near the coast, or a spaceport (in both time and space), or have access to a large helium balloon and camera, then what other demonstrable evidence is there to show Earth isn't flat? Well, given enough of an apparently "flat" space, two sticks, and a sunny day, we can indeed show that Earth is not flat. To put it another way, we can prove to the world that the world's nice and round just by sticking two sticks in the ground. Added to that, if we are careful enough with our measurements and assume Earth is a perfect sphere (which it isn't, but it's close enough), we can use this method to actually estimate its size.

Consider the surface of a sphere with two sticks driven into it at a right angle to the surface. If we get it right, the bases of those sticks will be slightly closer together than the tops because of the curvature of Earth. If the sticks were long enough and driven far enough in, they would meet at the center of the sphere. This fact must be taken into account when architects design large buildings and suspension bridges. For example, the Vehicle Assembly Building at the Kennedy Space Center is about 0.5 centimeters wider at the top than the bottom due to the curvature of Earth, and the tops of the Golden Gate Bridge towers are about 4.5 centimeters farther apart than the bases for the same reason [8,9].

Astronomy Saves the World

Two sticks in the ground are easier to deal with than huge buildings and bridges, but there are still several things that could be varied in this experiment: the altitude of the sun when we take the measurements, the distance separating the sticks, and the length of the sticks themselves.

As we move about on the surface of Earth, the distance of the sun above the horizon, at any particular time, changes. This fact, coupled with the rotation of Earth, is why we need time zones. New York is five time zones behind London. When people are having their breakfast bagel in New York, Londoners are perhaps enjoying a succulent sausage lunch and a pint in a nice pub. The sun is clearly at two different altitudes at the two different locations at the same absolute time. Regardless of eating arrangements, two equal-length sticks stuck in the ground with a clear view of the sun in London and New York (not an easy thing to get in either location) are going to cast shadows of different lengths. The sun is high in the sky in London (short shadow) but low in the sky in New York (long shadow).

Returning to trigonometry, it turns out that the ratio of the length of the stick to the length of the shadow, i.e., the length of the stick divided by the length of the shadow, gives us the tangent of the altitude of the sun at each location. This measured difference in the sun's altitude between London and New York (which should be about fifty degrees or 0.87 radians), plus the known separation of the two locations (about fifty-six hundred kilometers), allows us to use the relationship between an arc length (fifty-six hundred kilometers) and an angle (0.87 radians) to determine the radius of Earth, i.e., $5600/0.87 = 6400$ kilometers. The accepted *average* value today is 6370 kilometers, give or take the lumps and bumps.

The London to New York measurement is not exactly something that could have been made by someone two thousand years ago. Therefore, to make the measurements practical, things need to be scaled down. In fact, because the timings of the measurements need to be synchronized for observations made on just one day, it would really help if—in the absence of modern technology like radios, cell phones, and GPS—the two sticks could be seen from each other.

The accuracy to which Earth's radius can be determined depends on how well we can measure the relative heights of the sticks and their shadows. Consider two meter sticks placed only one hundred meters apart. If the sun is precisely over the head of the first stick, the stick won't cast a shadow. Due to the curvature of a spherical Earth with a 6370-kilometer radius, the shadow at the second stick will be 0.016 millimeters. This measurement faces several significant challenges. It assumes that both sticks can be driven into the ground at precise right angles to the ground itself, that the ground is perfectly "flat" when it may actually be slightly bumpy, that the sun passes directly overhead, that the observations can be precisely timed, and that we can easily measure 0.016 millimeters.

There are several solutions that can be used together to meet these challenges. The sticks can be taller, the sticks can be moved farther apart, and the measurements can be made at some time other than high noon. With meter sticks one kilometer apart, the effect (at noon) is now 0.16 millimeters. With the sun at an altitude of forty-five degrees to the first stick (three hours before noon), the effect is now 0.31 millimeters. On a really "flat" place, like very long beaches, the sticks could be placed up to five kilometers apart and still be seen from each other. If the measurements are both made two hours after sunrise, there will be a 3.14-millimeter difference in the shadow lengths, per meter length of the sticks.

Regardless of the chosen flat place, we are now presented with an effect that could be measured relatively easily, and this is the type of approach the Greek mathematician Eratosthenes (ca. 276–195 BC) used to estimate the radius of Earth around the year 240 BC. Eratosthenes was only about 10 percent off from the currently accepted value, and all he had was a water well in what is now

Aswan, Egypt, a trip eight hundred kilometers north to Alexandria, and a human brain[10]. The water well in Aswan was fortuitously positioned so that at noon on the summer solstice, there were no shadows on the wall of the well and sunlight could be seen at the bottom. In contrast, the sun in Alexandria—measured from the shadows cast at noon by tall objects on the summer solstice one year later—was found to be about 7.2 degrees away from being directly overhead. And from these observations, Eratosthenes calculated the circumference of Earth using basic trigonometry. The radius is a simple step from there.

The Closest Things Are Far Away

The closest astronomical object is the moon, the closest star is the sun, and the closest major galaxy is Andromeda. How do we know the moon is the closest of the three? Some basic observations of the moon give us some clues that ultimately lead to the answer. These observations can even be made without any specialized equipment.

With the exceptions of the occasional near-Earth asteroid or artificial satellites of Earth like the ISS or the *Hubble Space Telescope*, we won't see anything pass between Earth and the moon. About twice a year, some small areas of Earth experience a solar eclipse, which is when the moon passes directly between the sun and Earth. The moon either blocks out almost all the sun's light, casting a shadow onto parts of Earth, or blocks out some of the sun's light, making it look like something has taken a bite out of the sun.

Another noticeable fact is that the moon is regularly visible during both the day and night. If we note the time of moonrise or moonset today, that time will be almost an hour later tomorrow. If we see a moonrise at eleven o'clock at night, twelve days later, it will be rising at eleven o'clock in the morning. This simple observation tells us that the moon moves across the sky at a different rate than the sun. This motion is due to the orbital motion of the moon around Earth. At night, we can easily see the motion of the moon with respect to the other stars, because they all pass behind it, and it takes the moon 27.3 days to make a complete lap of the sky, i.e., to orbit Earth.

During this time, the bright parts of the moon go through phases, which tells us that the sun is casting light on the moon, just like it casts light on Earth. When the moon is close to the sun on the sky, we do not see it as easily. This is because the daytime sky is bright, and we see the dark side of the moon. The arrangement is sun–moon–Earth. When the moon is farthest away from the sun on the sky, we see the bright, sunlit side of the moon during the nighttime. The arrangement is sun–Earth–moon.

Compared to solar eclipses, lunar eclipses are seen far more often. This is when the moon is completely covered by Earth's shadow, and when they occur, they can be seen on the cloudless night side of Earth by anyone, including the ancient Greeks. If the length of Earth's shadow can be calculated, then we will be able to determine the distance to the moon. As long as we know the diameter of Earth, which, thanks once again to the Greeks, we have known for millennia, determining the length of Earth's shadow is rather straightforward.

From Earth, the sun has an angular diameter of thirty-one arcminutes (0.52 degrees). Take an opaque marble that is one centimeter in diameter, a golf ball that is 4.3 centimeters in diameter, or a basketball that is twenty-five centimeters in diameter, and see how far away they need to be placed so they have an angular diameter of thirty-one arcminutes. In the case of the marble, it will be at a distance of 1.1 meters. In the case of the golf ball, it will be 4.8 meters. The basketball will be twenty-seven meters away. At these distances—if held correctly—each object will totally eclipse the sun, meaning the lengths of their shadows must be about 1.1, 4.8, and 27 meters, respectively. All these results could have also been calculated using the simple angle relation of $\theta = s/r$. The ratio of the distance at which it subtends thirty-one arcminutes to its actual diameter is always 109. It is the same case for Earth. The length of its shadow must be 109 times its diameter, which would make it 1.4 million kilometers.

Since lunar eclipses prove that the moon falls completely into Earth's shadow, the moon must therefore be closer than 1.4 million kilometers. This might not seem that helpful, but in astronomy, any limit we can set is important. We now know that the sun must be farther away than 1.4 million kilometers and therefore bigger than the moon. In addition, when we come up with a way to find the distance to the moon more precisely, that result must be consistent with this 1.4-million-kilometer limit.

During a lunar eclipse, the moon moves through Earth's shadow, which has a cone shape. It starts as a circle the size of Earth and gets smaller and smaller until it reaches the end point (the apex). Knowing this, we can measure the angular size of Earth's shadow cone at the distance of the moon. Two obvious limits to the size of Earth's shadow cone are the diameter of Earth itself (the shadow cone can't be larger than that) and the actual diameter of the moon (the shadow cone must be larger than the moon's diameter to cause a total lunar eclipse). Again, these limits are useful, since they set basic boundaries on the answer.

There are at least three approaches that use Earth's shadow cone to constrain the distance to the moon.

The first method is the most direct: We simply stare at the moon as it proceeds through the shadow. We know that the angular diameter of the moon is a little over a half degree, so all we need to do is to count how many moons fit inside the shadow. However, this observation will depend on which *part* of the shadow the moon passes through. If the moon travels close to the edge of the shadow cone, it won't be in eclipse as long as when the moon travels right across the middle. This is because the moon's orbit around Earth is tilted, and it travels through different regions of Earth's shadow during different eclipses. Also, since the moon does not orbit Earth in a perfect circle, sometimes the moon is closer to Earth and moving faster, and sometimes it is farther away and moving slower. Nevertheless, after some careful staring, this method finds that 2.7 lunar diameters fit inside Earth's shadow cone.

The second method is based on what we know about the moon's orbit. It takes about 27.3 days for the moon to complete one orbit around Earth, which corresponds to about 0.55 degrees per hour. Therefore, we can time how long the moon spends in the full shadow of Earth. We can then determine the size of Earth's shadow at the distance of the moon (speed equals distance divided by time, distance equals speed times time). After carefully staring at multiple lunar eclipses, we'd find that an hour and forty minutes is the longest time for the moon to be entirely in Earth's shadow. This means the angular diameter of Earth's shadow, at the distance of the moon, is 1.4 degrees or 2.7 lunar diameters (an extra lunar diameter must be added to get the full width of the shadow).

The third method is a little trickier, but as the moon passes into Earth's shadow, the curvature of the shadow is projected onto the face of the moon. If we can determine the curvature of the shadow, we can also determine its radius, because smaller circles have tighter edges. The calculation can be a little complex, but it can also be used during a partial lunar eclipse as well as during the beginning and end of a total lunar eclipse. The measurement to be made is as follows. Take a line that joins the two points where Earth's shadow meets the edge of the moon. The position of the edge of the shadow's curve at the mid-point of that line will be at some distance from that line. In geometry, this distance is called the *sagitta*[11]. When the length of the first line drawn equals the diameter of the moon, the sagitta of Earth's shadow is about 10 percent of the moon's diameter. The calculation of the shadow's radius then uses well-known geometric features of circles and, if done with enough precision, will also result in a shadow size of 2.7 lunar diameters, or 1.4 degrees.

After a few minor mathematical and geometrical considerations (and some really good staring), it turns out that we can successfully determine the size of Earth's shadow, in terms of lunar diameters, at the distance of the moon. As lunar eclipses are quite common, these methods can be repeated over and over until we are happy we have a precise answer. This is exactly the type of evidence that makes for airtight results.

Let's now say the moon is at a distance of one lunar distance. At this distance, Earth's shadow is 2.7 times the diameter of the moon. If the moon were at the end of Earth's shadow, the size of Earth's shadow on the moon would be zero. Therefore, if the moon were pushed 2.7 times farther away from where it is now, to a distance of 3.7 lunar distances, it would actually be right at the end of Earth's shadow cone. Not "wedged" in the cone like a first scoop of ice cream but right at the end where Earth's shadow ends. From this, we can see that 109 Earth diameters is equal to 3.7 times the distance to the moon, i.e., the moon is at a distance of about thirty (109/3.7) Earth diameters. Consequently, because the diameter of Earth is known, the moon can be calculated to be at a distance of about 380,000 kilometers. The moon's orbit isn't perfectly circular, though, so this number varies by about 5 percent, but the Greeks knew this lunar distance over two thousand years ago. Regardless of the precision, compared to our everyday experiences, 380,000 kilometers is far away. And this is pretty much the closest astronomical object there is.

To help with the visualization of the distance between Earth and the moon, we can consider Figure 4.4 (which is not to scale). In it, we can see Earth and the moon as large and small circles, respectively, and their shadows extending to the right of them as triangles. The top example shows where the moon would be in relation to Earth if it were at the end of Earth's shadow (3.7 lunar distances) and no lunar eclipses occurred. The middle example shows how the moon would be in relation to Earth if it were wedged into Earth's shadow, like a scoop of ice cream in a cone.

If the moon were here, at 2.7 lunar distances, total lunar eclipses would be rare considering a perfect orbital alignment would be needed. The bottom example shows where the moon actually is in relation to Earth based on our observations of lunar eclipses. Knowing that there are 109 Earth diameters for every 3.7 lunar distances means calculating one lunar distance is trivial when we know the diameter of Earth.

The usefulness of the number 3.7 doesn't stop at determining the distance to the moon, though. It also allows us to calculate the actual size of the moon. These observations tell us that Earth's diameter is 3.7 times greater than the moon's diameter. Since we know Earth has a radius of about sixty-four hundred kilometers (thank you, Eratosthenes), the math tells us that the moon has a radius of about seventeen hundred kilometers. This can also be confirmed by remembering, once again, the simple angle formula $\theta = s/r$. The angular diameter of the moon (θ) is thirty-one arcminutes, and the distance to the moon r is 380,000 kilometers; therefore, the diameter of the moon, s, is about thirty-four hundred kilometers.

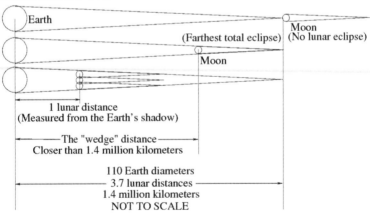

Figure 4.4: Calculating the lunar distances based on eclipse observations.

Astronomy Saves the World

It was Aristarchus of Samos (ca. 310–230 BC) that first documented calculating the distance to the moon. His efforts are quite remarkable, considering he didn't have a telescope. Aristarchus went on to postulate, perhaps for the first time in human history, that the sun was the center of our universe rather than Earth. He was also able to place the six then-known planets in their correct distance order from the sun. But Aristarchus was not able to get a good estimate for the average distance between Earth and the sun, This distance is known as the *astronomical unit* and it is named sensibly because, compared to the everyday experience of a human, it is a rather astronomically large number. It is from this number that we determine the distances to everything else in our universe, and seeing that we cannot physically measure it, it is quite important to calculate it as accurately as possible.

Recall the ratio of 109 we've seen a lot, the ratio of object diameter to shadow length? As with the shadow length of Earth, 109 is also the ratio of the distance to the sun to the diameter of the sun. However, knowing this doesn't help determine the distance to the sun. The sun could be relatively small and close (although not smaller and closer than the moon) or far away and large.

Rather than using the ratio of 109, Aristarchus attempted to determine the distance to the sun using the observed angle between the sun and the moon when the moon appeared to be 50 percent illuminated. However, at the time, this couldn't be measured accurately enough to get a good result. Consequently, he calculated the sun to be sixty-two times closer than it actually is. We would have to wait for the human race to become much more organized on a global scale before the actual distance to the sun could be determined accurately.

Nevertheless, over two thousand years ago, some really good staring got us to the point where we could easily demonstrate that Earth is roughly spherical with a radius of about sixty-four hundred kilometers and the moon is, on average, at a distance of about 380,000 kilometers. Neither of these facts makes us rely on any beliefs, and anyone with enough patience can reproduce the experiments described and see these truths firsthand.

What the discussions in this chapter have demonstrated is that anyone with an interest in the most fundamental questions in astronomy must have an appreciation of basic trigonometry and that gaining that appreciation is not outside the realm of possibility. It's just the second hand on a clock! What is more, trigonometry is a skill that has many more very useful applications far outside astronomy, so it would be great for the world if everyone were able to carry out basic trigonometric calculations and grasp their importance.

Compared to the distance from Earth to the moon, the size of our universe is such that the distance to the next nearest major astronomical object (one of the other planets or the sun itself) is a huge step up in scale. But we will be able to apply the same trigonometry covered here to figure these scales out. However, at these next scales, the accuracies in distance depend on the precision with which we can make measurements. For this, we must turn to technologies that help us see our universe more clearly. We can use small telescopes, most of which can cost less than one hundred dollars, to massively expand our understanding of the scale of our universe. And as we've stared deeper and deeper into the vast darkness with newer and newer technologies, we have uncovered facts about our universe that we never dared dream before.

Chapter 5: The Scale of the Problem

How far is the sun? This is another example of a simple question in astronomy, but it doesn't have an easy to find answer. This distance is the astronomical unit we've previously seen, and we will use this number to quantify the distances to everything in our universe. It will show us how Earth and the sun measure up on these scales, and it will begin to provide us with some much-needed cosmic humility.

One of the problems we have when trying to understand the scale of our universe is that it is completely outside our everyday experiences. It is ninety-three million miles to the sun. To try to put that into some perspective, consider how far a typical person travels on a typical day. About ten miles, probably. Not many people will have a one-mile commute to work. Not many people will have a one-hundred-mile commute either. So, barring any extenuating circumstances as well as for the ease of calculations, we will stick to this number for now.

The methods applied to this simple "how far" type of thought exercise are useful for a number of reasons, and they have a particular name. We call them *Fermi* problems, after being popularized by the famous atomic physicist Enrico Fermi (1901–1954).

The atomic physics problems that Enrico Fermi solved are—to say the least—complex, but the Fermi method is actually quite straightforward. It is essentially the *Goldilocks principle*. The *Goldilocks zone* around a star is frequently mentioned in astronomy, as this is where the temperature is not too hot and not too cold for liquid water to possibly be present on the surface of a rocky planet. The Fermi method is similar.

When we estimate a number, we should ask ourselves whether it is likely way too high, way too low, or about right. Is it *reasonable?* Our estimate doesn't have to be exact. It just has to be close enough to the truth to be useful. In addition, it doesn't have to be a number that we can necessarily grasp the scale of, so the Fermi method is particularly useful in astronomy.

Some things are obvious because they are part of the everyday human experience. What is the average height of a human? One centimeter is unreasonable, ten meters is unreasonable, but two meters is about right. Fermi problems can be straightforward or they can be relatively difficult. They are always easy to come up with, though, which makes them a favorite of professors. For example, how many cubic meters of water are there on Earth, and how many liters of freshwater are there on Earth *per person*?

There's No Harm in Checking

In today's world, many of us have continuous access to the Internet, where both reason and irrationality are demonstrated and put to the test. The Internet is littered with contradictions. It supports the truth and lies, good deeds and bad deeds, useful tools and useless playthings. However, on the whole, it seems to be leading our species toward a more productive future. Truths generally outweigh the lies, the good deeds eventually beat out the bad, and the useless playthings get old quickly and are left behind.

One incredibly powerful feature of the Internet is as a fact-checking tool, and students notoriously use it to help complete their homework, especially if they've been assigned a bunch of Fermi problems. Professors also use the Internet to check how much of a student's work might not be original. There are websites dedicated to fact-checking politicians, fact-checking urban legends, and even fact-checking the Internet itself[1,2,3]. The word "Google" is now a verb, and the phrase "Let me Google that for you" is used when the questioner should have Googled the answer to their question before asking others[4].

Fact-finding is another very useful feature of the Internet, and one that should be used when we come up with an idea[5]. In a world of seven billion people, and being a member of a species that perhaps has had a total of one hundred billion people throughout time, that idea, while entirely original to the person thinking it, may not be original at all[6]. So, if one day we realize that we can determine the total volume of a strange shape by submerging it in water and measuring how much the water rises, we might want to check the Internet before we run around our hometowns shouting, "Eureka!"

At the beginning of this chapter, I estimated (and I did this *without* looking at the Internet first) that the average distance that someone may travel in a day is ten miles. The Internet actually allows us to fact-check it! If I were to type in *how far does the average person travel per day*, Google would return a link, a little way down the first page, to the Federal Highway Administration run by the U.S. Department of Transportation[7]. On that website is a table of statistics from a report called "Our Nation's Highways." In that table are the average miles driven annually by someone around the year 2000: 13,476 miles. Today, that number is probably different, but not by much. If we take that number and divide it by the number of days in a year, we get thirty-seven, which is the estimated average number of miles driven per day by someone in the United States.

The United States is a relatively large, sprawling country. This could mean it has a higher than average daily distance traveled compared to most other countries, and there are plenty of people who don't drive. As a consequence, thirty-seven miles is likely close to a global upper limit, and the originally estimated ten miles per day isn't going to be far off the mark. Therefore, it is a reasonable estimate. But what does all this have to do with astronomy?

More Mathematics

Each time we skip from one, to ten, and then to one hundred, which we are forced to do a lot in astronomy because of the huge numbers involved, we are changing our scale by an *order of magnitude*. We can express these orders of magnitude by simply referring to the relative number of zeroes after the one. One hundred is two orders of magnitude greater than one, one thousand is three orders of magnitude greater than one, and ten thousand is three orders of magnitude greater than ten.

Like the trigonometric functions, there are also functions that help express numbers when they grow quickly. These are called *logarithmic* functions. When dealing with orders of magnitude, where the next order is ten times bigger or smaller than the previous one, we use a simple logarithmic function. The logarithm of one is zero, the logarithm of ten is one, and the logarithm of one hundred is two. To find the logarithm, or "log" for short, we count the number of zeroes after the one.

The opposite way of expressing the log of a number is to raise ten to the power of that logarithm. For example:

- For one zero after the one, we get log (10) = 1, 10^1 = 10.
- For two zeroes after the one, we get log (100) = 2, 10^2 = 10 x 10 = 100.
- For three zeroes after the one, we get log (1000) = 3, 10^3 = 10 x 10 x 10 = 1000.
- For four zeroes after the one, we get log (10,000) = 4, 10^4 = 10 x 10 x 10 x 10 = 10,000.

In this case, because we are always dealing with the powers in terms of ten, we say that this logarithm has a "base" of ten, but we can also choose to use other base numbers when it becomes convenient. The number of permutations for a given number of trials in a binary distribution (on or off, up or down, one or zero) can be described using the base of two. This is what today's computers typically use to store units of information (bits). Instead of the base of ten series going 1, 10, 100, 1000, 10,000, etc., the base two series goes, 1, 2, 4, 8, 16, etc.

When we consider our experiences in everyday life, we find that we are used to dealing with things on scales of a few orders of magnitude. We can readily give the number of millimeters in a meter and the number of meters in a kilometer, but we aren't so quick to give the number of millimeters in a kilometer (one million). We don't mind walking one mile, (which is, in England, about the average distance to the pub), we'd probably walk ten miles without too much fuss, but very few of us are interested in walking one hundred miles. We understand how long a year is, try to ignore how long a decade is, and get uncomfortable at the thought of one hundred years, possibly because only very few of us, for the time being, will live that long.

Ordinarily, we deal with about two or three orders of magnitude. But our universe doesn't give us that luxury. It blows the typical human experience of scale completely out of the water, into orbit, and beyond. Simply comprehending the scale of our universe becomes a problem in and of itself. One conceptual problem of living in such a vast volume like our universe is that there is ample room, and time, for improbable things to happen quite regularly. Even though stars explode very rarely, throughout our universe, thousands of stars are exploding every minute.

On the scale of our everyday experiences on Earth, about ten miles, getting to grips with the actual size of Earth isn't too difficult. It is a few orders of magnitude greater than our normal experience. When we take a flight traveling five hundred miles per hour from one destination to another, we really do get to experience the scale of Earth. The distance to the moon, on the other hand, is an order of magnitude greater again, i.e., about four orders of magnitude greater than the everyday experience. Most people can still probably appreciate this type of distance. The distance to the sun is where we are blown out of the water. It's about six orders of magnitude, or about one million times, greater than our everyday experience.

With distances like these, astronomy is once again making us learn a little bit more mathematics. It helps us make quick decisions on reasonable numbers, and it compels us to become familiar with scales that are outside our normal experience. In doing so, it turns our attention away from our self-inflicted Earth-based troubles and gets us to appreciate what really is the big picture. This is something I wish everyone got to experience. Perhaps, if this order of magnitude understanding were more widespread, then the people making budgetary decisions would see why saving a few million dollars here and there for a trillion-dollar deficit is fatuous.

Accuracy and Precision

In astronomy, words similar to "about" are often used when referring to the value of a number, i.e., the radius of Earth is about sixty-four hundred kilometers. This is done entirely on purpose, and there are some subtleties here that are worth discussing.

When "about" is used, it does not mean that we don't know what the answer is. It merely means that the actual number is close to the number presented, not known *precisely*, or we do know the number really well and the answer has some details that are not relevant in some cases. Remember, all we can do in astronomy is stare, so some numbers are really difficult to know really well.

The radius of Earth is a good example. At Earth's equator, the radius is 6,378 kilometers. At the Earth's poles, the radius is 6,357 kilometers. The numbers aren't the same, because Earth isn't a perfect sphere—it is slightly bulgy around the waist because it is spinning. In addition, there are other bumps and troughs that make it an irregular shape that we call a *geoid*. Consequently, we say that the radius of Earth is about sixty-four hundred kilometers. We purposefully remove some of the details of many different correct answers so that one number covers most cases. In fact, through most of this book, the numbers presented will only give the first two digits of the answer. They are the only two *significant* figures. This is to keep things as simple as possible while also being accurate.

The difference between accuracy and precision is important. They are most definitely not the same thing, but the difference can be a little subtle. "Accurate" means that the quoted values actually encompass the correct answer. The value of sixty-four hundred kilometers for Earth's radius is accurate, but as the implied range of possible values goes from 6,350 to 6,449, it is not very precise. "Precision" means that the range of quoted values is small. This indicates that we know the answer. But, it is possible to be precise and inaccurate.

Let's consider an example. Imagine the actual number of apples growing on a particular tree is 437. This is known because several people counted them all several times, and they all got the same answer. Now we come along and, for whatever reason, we have got thirty seconds to count all the apples on the tree. We don't have enough time to count them all individually. Flashing back to the Fermi method, we count forty apples in what appears to be one-tenth of the tree and estimate that there are four hundred apples. We then realize that some sections may actually have one or two more or less apples, so we estimate there are actually between 350 and 450 apples on the tree. We are accurate, but not very precise.

Let's now imagine we have more time to count the apples. We count the apples on the top half of the tree and find 261, so we estimate there are 522 apples total. We think we might have miscounted by ten apples in the top half, so we estimate there are between 502 and 542 apples. This is a more precise number of apples, with a spread of only forty apples rather than one hundred, but it's wrong. One explanation is that there is no reason to assume the apples are distributed uniformly throughout the tree.

A classic example to demonstrate accuracy and precision is with the hitting of a target. Figure 5.1 demonstrates the difference between the two. In astronomy, it is incredibly difficult to be precise, and so we are very cautious about how we present numbers. Ideally, we'd like to be both accurate and precise. Since there is no benefit in being inaccurate and precise (being wrong but sure the answer is right), we generally take the less risky, more pessimistic, approach and favor lower precision but higher accuracy (being right but not knowing by how much). Yes, we generally present approximate numbers on purpose. However, approximate numbers are not good enough when we send spacecraft on deep-space missions. In these cases, we must be both accurate and precise for the mission to be a success, so how have we been able to establish accurate and precise distances to the other objects in our solar system if all we can do is stare?

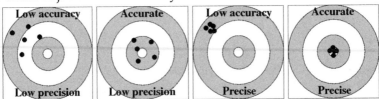

Figure 5.1: Target practice to show the differences between accuracy and precision.

Parallax

When determining the approximate size of Earth, we can use two different viewpoints of the sun by placing two sticks some known distance apart. With enough care, the lengths of the shadows can be used to determine the angle between those sticks, as if they had been extended all the way to the center of Earth. This leaves the radius of Earth as the calculable long axis of a triangle. This same method can be turned around so that the long axis of the triangle is now the distance to objects on the sky.

When traveling by whatever method (walking, biking, driving, sailing, flying, falling, launching, or orbiting), we notice the relative positions of the objects around us appear to change. Anyone that has been out in the sun on a particularly hot day will have used this effect when purposefully moving into the shade. This same effect can be used to estimate how far away something is.

Let's just consider traveling along a simple road for now. As we pass by objects, they appear to move through an angle. Objects that are close to the road, like lampposts, apparently move through a large angle. Objects that are far away, like mountains in the distance, apparently move through only small (even unnoticeable) angles. There is an obvious relationship between the apparent angle through which something appears to move and the distance from which we observe it (see Figure 5.2).

This distance-dependent change in apparent position as we view something from two different positions is called *parallax*. We can easily demonstrate the effect to ourselves by holding up a finger at arm's length. The apparent position of the finger will change, with respect to the background, as we alternate looking at it through only the left eye and then only the right. The apparent change in its position is greater when the finger is moved closer to the eyes. This effect gives us depth perception, i.e., the ability to get out of the way of something heading toward us, hopefully with enough time to spare.

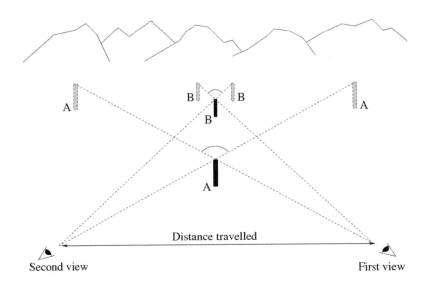

Figure 5.2: The parallax effect from nearby stick A and farther stick B. Stick A has a larger shift in apparent position against the background mountains because it is closer. The parallax angle is larger. Stick B has a smaller shift in apparent position because it is farther. The parallax angle is smaller.

When estimating distances using the parallax method, the actual size of the base of a triangle is going to be very important for the achievable precision. For the winking eye finger experiment, a baseline of about six or seven centimeters was used (the typical distance between your eyes) while the finger was held at a distance of less than about sixty or seventy centimeters (the typical length of an arm). But for measuring the distances to astronomical objects, these baselines must get much larger, because the relative distances are so much greater. There is a tradeoff between the distances to the objects, the baselines needed, and the precision of our instruments.

Thousands of years ago, the only instrument available to investigate our universe was a pair of human eyes. Therefore, the parallax effect—while known at the time—would not have been something easily noticeable on the sky, even for something really close. A telescope is almost essential in order to make these measurements, but Galileo Galilei (1564–1642) only began his telescope-enabled observations about four hundred years ago. That didn't stop the ancient Greeks, Hipparchus in particular, from trying to use the parallax method for astronomy.

Consider the moon as it plods across the sky. As it orbits around Earth, it is continuously passing in front of stars. Occasionally, the moon, viewed from certain positions on Earth, will pass in front of a planet. Jupiter is a good example for this. Like the left–right eye winking effect, parallax will make Jupiter appear to have a different offset from the position of the moon, depending on the location of the observer. This also means the moon will pass in front of Jupiter at two slightly different times from each location. The size of the observed Jovian spatial offset and the distance between the two observing sites allows the distance to the moon to be calculated using $\theta = s/r$.

As we saw earlier, Aristarchus got a pretty good estimate for the distance to the moon but couldn't get an accurate distance to the sun. He was attempting to apply more trigonometry by waiting until the moon was thought to be at a right angle between Earth and the sun. If a right angle triangle is present, and we know the length of one of the sides (the distance to the moon), only the size of a second angle is needed in order to determine the lengths of the remaining sides, the distance to the sun. Aristarchus thought that this second angle was eighty-seven degrees, which put the sun at $\tan(87) = 19$ lunar distances (seven million kilometers). The angle he tried to measure is actually 89.85 degrees, making the sun $\tan(89.85) = 382$ lunar distances away (150 million kilometers). A measurement error of "only" three degrees led to an answer that was wrong by over an order of magnitude! This is the way the tangent function behaves for angles approaching ninety degrees: a large punishment for a small error.

Astronomy Saves the World

There is almost no uncertainty in the angles involved during a total solar eclipse. If you happen to be in the right place at the right time, the angle between the moon and sun is precisely zero. So, like determining the distance to the moon from a lunar eclipse, can we determine the distance to the sun using a similar method?

Hipparchus realized the parallax method could, in principle, be used to determine the distance to the sun. In practice, it is not so easy. The scattering of the sun's light by Earth's atmosphere makes for a beautiful blue sky during the day, and orange sky during dawn and dusk. However, this scattering *almost* completely stops us from seeing any other stars and planets during the day[8]. It is possible to pick out the bright planets and stars during the day with a telescope, but even those objects would be lost in the glare of the sun. But what if the moon eclipses the sun?

During a total solar eclipse, the moon blocks out direct sunlight, and the stars can be seen. So is it possible to measure the parallax of these background stars during a total solar eclipse? At best, the width of the moon's shadow on the surface of Earth is only about two hundred kilometers. This means that if two observers were placed right at the edges of the shadow, the parallax of the sun against the background stars would be very small (0.3 arcseconds). Not only would such accuracy be beyond the abilities of the ancient Greeks, and even beyond the abilities of some small telescopes today, but also the observers would need to know ahead of time where the very edges of the eclipse shadow would be. These things make the total solar eclipse technique very challenging.

Hipparchus, no doubt keen to use the parallax method to determine the distance of something, eventually used solar eclipses to independently verify the Aristarchus moon distance. During a partial solar eclipse, only a fraction of the sun's disk is blocked by the moon. In this case, we would know the distance between two observing points (one in total eclipse, one in partial eclipse) and an estimated angle based on the amount of the solar disk that the moon is blocking. Using this method, Hipparchus estimated a lunar distance that was only 10 percent larger than that estimated by Aristarchus. That's pretty good, in astronomical terms.

The Wandering Stars

After the progress the Greeks made in figuring out the size of Earth and the distance to the moon, an accurate distance to the sun was a long time coming. This was simply due to the precision needed to make the necessary observations, even when a semi-reliable method was figured out. However, Aristarchus had still taken a few steps in the right direction when he placed the planets in the correct distance order from the sun. It turns out that the first key to calculating the distance to the sun is not in finding the distance directly but in understanding the relative distances to the "inner" planets that are closer to the sun than Earth.

Humans enjoy looking for patterns, even when there aren't any. The face on Mars is a prime example. The technical term for this pattern-finding effect is *apophenia*. The constellations are another good example. Stars appear to show patterns similar to idols in particular cultures. Most likely, ancient idols were forced onto patterns in the stars or invented to fit the stars. The ancient Greeks saw the belt of the great hunter Orion in one pattern of stars, while the ancient Egyptians saw the crown of the god Sah in the same three stars[9]. Not only did these two different cultures have two different mythologies to build patterns from, but they were also able to see different parts of the sky and, therefore, different stars. Egypt is a lot farther south than Greece, so the Egyptians, particularly those in Luxor or Aswan, are able to see stars not seen in Greece.

Our collective apophenia primes us to notice when something is different. Over a period of a few hours, it becomes apparent that the moon moves differently from the stars. Over a period of a few weeks, it becomes apparent that the sun does something different too—it rises and sets at different points on the horizon, not always due east and west. Over periods greater than a few weeks, it becomes apparent that some star-like objects "wander" in and out of constellations. More curious is that, every once in a while, the objects will turn around and go retrograde (backward) on the sky.

To the unaided eye, there are five of these "wandering stars": Mercury, Venus, Mars, Jupiter, and Saturn. A telescope is needed to see Uranus and Neptune. In fact, the word "planet" means "wandering star" in ancient Greek, and when these five planets are combined with the sun and the moon, we end up with seven handy names for days of the week[10]. Astronomy is ingrained in human culture.

Each visible planet moves with a different speed across the sky and therefore spends a different amount of time in a constellation. Mercury passes through the constellation Gemini in about two weeks, Venus in about three weeks, Mars in about six weeks, Jupiter in about three or four months, and Saturn in six or seven months. The sun takes about four weeks due to the orbit of Earth. Mercury and Venus go from one side of the sun to the other. Mars, Jupiter, and Saturn "wiggle" back and forth on the sky, regularly performing their retrograde pirouettes. The apparent backward motion of these planets happens when Earth overtakes them on its smaller, faster orbit around the sun.

For most of human history, our species believed that Earth was the center of our universe. However, as we shall see, once this idea began to buckle, it was a simple matter to arrange the planets in the correct order, with the sun (known in Greek mythology as Helios) at the center. This gives us the *heliocentric model*. Mercury is fastest and appears on one side of the sun and then the other. It is closest to the sun. Venus does the same, but much more slowly, and it separates itself farther from the sun on the sky than Mercury. It is second closest. Mercury and Venus are called *inferior* planets, because they are closer to the sun than Earth.

Based on the apparent motion of the sun, Earth comes next. This also provides an explanation for the retrograde motion of the three remaining planets that are seen with the naked eye. They are *superior* planets, because they are farther from the sun than Earth is. Mars is the fastest of the three, followed by Jupiter, and then Saturn. It is their observed speeds across the sky that determine their distance order from the sun. The farther away they are, the slower they travel.

Observations and reasoning like this would have enabled Aristarchus to place the planets in distance order, as he did. However, back then, it was too much for popular culture to displace Earth from the center of our universe and have it hurtling through space. Even Aristotle wasn't convinced.

In the meantime, many cultures dabbled with heliocentrism, but it wasn't until around eighteen hundred years later that Aristarchus's idea was thoroughly resurrected by the European astronomer Nicolaus Copernicus (1473–1543). Unlike Aristarchus, Copernicus didn't have an Aristotelian figure to dismiss his sun-centered idea. Instead, his idea had the venerable honor of upsetting the Catholic Church. This self-proclaimed moralistic authority compelled some of its followers to carry out horrific punishments on those who dared *think* about reasonable explanations for the mechanics of our universe, especially those that contradicted explanations prescribed by their particular dogmatic paradigm. This hindered the spread of the heliocentric idea.

Regardless of theological pressures, the retrograde motion of the superior planets still posed a significant problem for all proposed solutions that kept Earth at the center of our universe. These are the *Ptolemaic* (geocentric) models named after the Alexandrian Roman citizen Claudius Ptolemy (cr. 100–168 AD). Copernicus, on the other hand, placed the sun at the center of the solar system with Earth and the five other planets moving in perfect circles around it. Even if it couldn't explain everything, the heliocentric Copernican model is a much better explanation of the observed motions of the planets than the geocentric Ptolemaic model. However, as the precision of the observations increased, the inconsistencies between the predictions of both models and the recorded positions of the planets became more of a problem.

Tycho Brahe (1546–1601) was a renowned Danish astronomer who could record the positions of celestial objects on the sky, including the planets, with a new level of accuracy and precision. Brahe's accuracy with celestial positions is incredibly important but frequently overshadowed by the interesting social life he probably led (he wore a prosthetic nose, for example). The more accuracy with which the motions of the planets were mapped, the more problems the simple perfect-orbit Copernican model had. Brahe stuck with the church-pleasing geocentric model of the sun traveling around Earth. However, in a tip-of-the-hat to Copernicus, he modified the classic Ptolemaic model to allow the five visible planets to travel in uniform circular orbits around the sun.

The problems for Brahe's model didn't come from the Church, or from the motions of the planets—it came from the observed paths of comets. The motions of these wandering objects were inconsistent with the motion of the planets and entirely inconsistent with uniform circular motion around the sun. They were definitely going to fly by the sun, but there was no way that they could be on circular orbits.

In 1600, Johannes Kepler (1571–1630) began assisting Brahe in his work. While there was contemporaneous evidence for the heliocentric model based on the phases of Venus as observed by Galileo, it wasn't until the year after Brahe's death that Kepler reapplied the heliocentric model of Copernicus to all Brahe's data. He did this without having to worry about offending any personal beliefs that Brahe may have had. Instead of the motions of the planets being perfect circles, Kepler posited, why couldn't they be ellipses? It turns out that not only *can* the orbits of the planets be elliptical; they *are* elliptical. Unlike circles, ellipses have a few complications that are worth pointing out.

A circle is one specific kind of ellipse, but an ellipse does not have a constant radius. It has a range of radii going from the largest (shown in Figure 5.3 by arrow A) to the smallest (shown by arrow B). The largest axis stretches across the stretched direction of the ellipse, and the smallest axis is squished in the squashed direction. Arrow A is called the *semi-major* axis, because it is only half the total distance across the largest (major) part of the ellipse. Arrow B is called the *semi-minor* axis, because it is only half the total distance across the smallest (minor) part of the ellipse. If A were equal to B, we'd have a circle. C marks the center of the ellipse. F1 and F2 are two focus points. They are the points around which objects on elliptical paths rotate.

The closer the object is to a focal point, the more influence that point has on the path of the object. Imagine the dashed line is a piece of string, and the string is tied to pins at points F1 and F2. P is a pen or pencil. If the string is kept taut, then as the pen is moved around, it will trace out the elliptical path. The total length of the string never changes, so the distance from F1 to P added to the distance from P to F2 is always constant.

Kepler realized that instead of the P being a pen or pencil, perhaps P could be a planet and F1 could be the sun (or F2; it doesn't matter as long as there is only one sun). Instead of the planets moving in perfectly uniform circular motions, as in classical Copernican heliocentrism, they are following elliptical orbits. The realization of elliptical motions for the planets was a step forward in thinking, which answered some of the questions that had held back heliocentrism for so long.

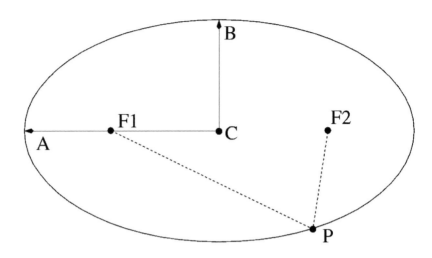

Figure 5.3: The basic details of an ellipse as they pertain to orbits. (A) is the semi-major axis. (B) is the semi-minor axis. (C) is the center. (P) is the planet. (F1) and (F2) are the foci. The sun would be at one of these foci. The sum of the two dotted lines is constant no matter where P is on the ellipse.

The construction of the elliptical model was so successful that it became Kepler's first law: planets follow elliptical paths. We can assign a parameter to describe how un-circular an ellipse can be: *eccentricity*, and we can even give it a letter: e. For a perfect circle, $e = 0$. For Venus, $e = 0.0068$. Earth has $e = 0.0167$. This demonstrates why the high-accuracy observations of Brahe caused some problems for Copernican heliocentrism. Planets in our solar system typically have eccentricities of less than 10 percent (Mercury is the exception), but Brahe was able to achieve the precision required to detect this effect. Kepler's first law also explains the motions of comets; they follow ellipses too, albeit with very high eccentricities. Kepler went on to address several other obvious things about the planets, like how far away they are and how long they take to orbit the sun.

Earth takes one year to complete a single lap around the sun, and its relative distance to the sun is one. One *astronomical* unit. These two things by themselves don't seem that helpful, especially as we need to know what an astronomical unit actually is in kilometers if we are to successfully navigate space missions through the solar system. Setting both these numbers to one allows us to find the *relative* distances and *relative* periods of the other planets as compared to Earth. Calculating the relative distances to Mercury and Venus relies on (once again) simple trigonometry. All we need to know is the largest angle each inferior planet has with respect to the sun.

As an inferior planet orbits around the sun, on the sky, it appears to get farther and farther away, stop, and then go back. We can measure the angle between the planet and the sun at the point where the planet appears to stop and simply take the sine of that angle. This would give us the distance of that planet in astronomical units, as shown in Figure 5.4.

With the sun at the center, Earth and an inferior planet have orbits given by two circles. The ellipticities of the orbits are being ignored in this case. The biggest circle represents Earth, the smaller one Venus or Mercury. When the largest observed angle (A) is reached by one of the inferior planets, the line to that planet from Earth forms a right angle with the line from that planet to the sun. We have a right-angled triangle.

The distance on the long side of the triangle is the Earth–sun distance (one astronomical unit). The angle between the sun and the planet provides the last piece of information needed to get the relative distance. In this case, it is simply the sine of that measured angle. For Mercury, that angle is twenty-three degrees, on average. This makes the relative distance between Mercury and the sun about $\sin(23) = 0.39$ astronomical units. For Venus, that angle is forty-six degrees, giving a relative distance of about $\sin(46) = 0.72$ astronomical units.

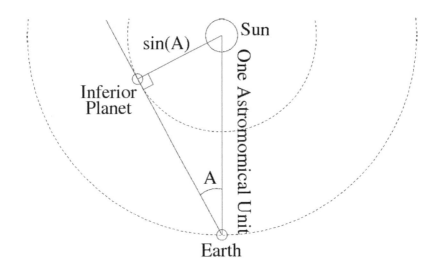

Figure 5.4: The trigonometry needed to determine the relative distances of the inner planets. Angle (A) can be observed so that the distance sin(A) can be calculated in astronomical units.

We can make these observations of Venus and Mercury without a telescope, but we need a lot of patience. Alternatively, when a telescope is available, we can determine when an inferior planet reaches an angle of ninety degrees by looking at its phase. When the planet has reached exactly one-half phase (half its disk is lit), we know it is in the right position for the sun angle to be measured and for the relative distance to be calculated.

The distances to the superior planets—those farther from the sun than Earth—are a little trickier to calculate. They don't show phases, because we always see their fully lit (day) side. To determine these distances, we must go back to Brahe's precision position measurements and the heliocentric model. It turns out that trigonometry is the key again, and with careful observations, the distances of all the planets from the sun can be determined in terms of astronomical units.

The relative distances of inferior planets from the sun are not the only observations we can make without the aid of advanced instrumentation. The time it takes a planet to go around the sun (its period) isn't too tricky to figure out either. One complication is properly accounting for Earth's own orbital motion. Even though Venus's orbital period is about 225 days, it takes 584 days to return to the same relative position to the sun. This is known as its *synodic* period. In that time, Earth has gone around the sun more than one and one-half times. Mars has an orbital period of about 687 days, but it takes 780 days to return to the same relative position from the sun.

Planets closer to the sun make complete orbits *more* frequently than Earth does. The true orbital period of an inferior planet is found by adding the two orbital frequencies together, i.e., the orbital frequency of the planet equals the orbital frequency of Earth plus the synodic frequency (frequency is one divided by the period). Planets farther from the sun make complete orbits less frequently than Earth does, so the actual period of a superior planet is found by subtracting the two frequencies from one another, i.e., the frequency of the planet equals the frequency of Earth minus the synodic frequency. Brahe's high-accuracy observations made it possible for Kepler to determine accurate and precise periods for the five observable planets.

Another perk of the high-accuracy observations of planets is that their orbital velocities can be seen to change as their orbits progress. These observations provoked Kepler to establish his second law. If we combine how far the planet moves in a given amount of time (a length) and its distance from the sun (another length), we find the result is a constant no matter where on the elliptical path the object is. Since we are considering two dimensions of length, which give an area, Kepler's second law states that planets "sweep out" equal areas on their elliptical orbits in equal times. Kepler's second law is a result of gravitation; as something gets closer to a massive object, it speeds up; as it gets farther away, it slows down. However, this wasn't known at the time, because the father of gravitational theory, Isaac Newton, wasn't born until twelve years after Kepler's death.

Kepler's third law demonstrates a classic astronomer "trick" of seeing whether there are any underlying relationships in the data. With the relative distances and periods of the planets, Kepler did indeed find a relationship. The square of the period in years (time times time) is equal to the cubed semi-major axis length in astronomical units (length times length times length), or $P^2 = a^3$.

When relationships like this are found, they are entirely empirical; they are purely based on observation and do not come with any explanation as to *why* these rules are apparently followed. An observed relation doesn't tell us much, if anything, about the underlying physics. An apparent correlation does not guarantee two things are related, but we now know that Kepler's laws are explicable through Newton's gravitational theory. Once Newton realized that gravity explains the motions of the planets, moons, and comets in terms of orbits around the sun, the proof of Kepler's third law was quite simple.

The Distance to the Sun

It is probably worth pausing to remind ourselves that we are trying to determine the distance to the sun (the astronomical unit), and the point of this whole chapter is to begin to demonstrate that our universe is really, really big.

We now know that the parallax effect can be used to determine the distance to an object. We also know that there are two foreground objects on the way to the sun, Mercury and Venus, and how many astronomical units they are from the sun. The sun can be used as the background object (the faraway mountains) and the planets as the objects we calculate the distances to (the nearby sticks). Once we know the absolute distances to the inferior planets, we will know everything we need to know to calculate the absolute distance to the sun.

Parallax allows us to find the absolute distance to Venus as it passes in front of the sun. We can then simply divide that distance by 0.28 (the number of astronomical units Venus is from Earth) to get the absolute distance of Earth from the sun. All we need to do is wait for Venus to pass in front of the sun and observe the event from two different places on Earth. These two places will form the base of our parallactic triangle from which to make the distance estimate.

Due to its orbital tilt, Venus can "miss" the 0.52-degree solar disc from our point of view on Earth. This means that a transit across the solar disk does not occur at the synodic frequency and makes a transit event quite rare. It probably comes as no surprise that Kepler was the one who figured out how to calculate when these transits of Venus might actually occur. After all, he had the data and the laws to make the predictions. So, in 1627, he predicted a Venus transit would occur in 1631 with a follow-up near transit eight years later. Kepler was slightly off in his calculations, though, and 1631 was a near transit. It wasn't until 1639 when the transit of Venus was first observed.

Jeremiah Horrocks (1618–1641) and William Crabtree (1610–1644) independently made the observations of the 1639 transit using specially modified telescopes and updates to Kepler's calculations. Unfortunately, they did not have the precision and baseline required to get an accurate estimate of the size of the solar system, but they were in the ballpark and showed that the distance to the sun was at least a massive ninety-five million kilometers.

In the mid-seventeenth century, the scientific, philosophical, and theological implications for determining the scale of the solar system were very high. Following the work by Horrocks and Crabtree, multiple governments invested in attempts to record the Venus transits predicted for the eighteenth and nineteenth centuries from the widest baselines possible. These global adventures, which included the likes of James Cook (1728–1779), had other merits too, such as testing navigational methods and finding new land in partially mapped oceans. Astronomy once again paved the path of human progress.

The parallactic Venus–sun transit method in and of itself is quite simple and was realized by the comet guy, Edmund Halley. From two different observing points, Venus appears to transit across two different segments of the solar disk. One path is apparently closer to the equator of the sun than the other, and that difference is the parallactic angle.

Let's take the June 5 and 6, 2012 transit of Venus and see what the difference was from two widely spaced observing locations like Sydney and Tokyo. In this case, the baseline is seventy-eight hundred kilometers. Observed from Tokyo, Venus passed across the disk of the sun 1.3 arcminutes (0.022 degrees, 0.00038 radians) closer to the solar equator than observed from Sydney. However, by applying some simple trigonometry, we find that the distance to Venus during a solar transit is forty-two million kilometers. When divided by the *relative* distance to Venus (1 minus 0.72 astronomical units), this makes the distance to the sun a whopping 150 million kilometers.

When averaged over all the previous efforts, the distance estimates for the sun did indeed come out to be about 150 million kilometers. As a comparison to the observable offset, the angular diameter of Venus itself during a transit is about one arcminute. This makes the observations necessary a challenge. As an alternative, the relative timing of the transit can be used. It takes Venus longer to transit across a different part of the disk, because the disk has a varying width (maximum at the equator, zero at either of the poles). The transit observed from Tokyo lasted about nine minutes longer, which means Venus transited closer to the solar equator than observed from Sydney. Those nine minutes correspond to only a small percentage difference in the timing when the total length of the transit was about six hours, but knowing the approximate position of the transit on the disk and the difference in the transit timing still allows the parallax angle and distance to be calculated. The timing method comes with its own complications, though. The atmospheres of Earth and Venus can distort the path of the transit shadow and make it look as if the transit has started or ended when it really hasn't.

Nowadays, technology has overcome these early complications, and we can get remarkably accurate distances to most objects in the solar system by bouncing radio signals off them. Because we know these signals travel at the speed of light, timing how long the radio echo takes to come back is all we need for the distance calculations. So, regardless of how difficult and complicated the data collection and analysis needed to be back then, all the efforts that went into calculating the distance to the sun were not in vain, because the numbers that were determined then, 150 million kilometers, are the numbers that are still used today to define one astronomical unit.

The astronomical unit is an astronomical number, so it is no coincidence that the term "astronomical" is used as a metaphor for something really, really big and/or very important. In astronomy, the astronomical unit is both. This distance also forms the foundation of our distance scale for the rest of our universe.

Let's deal with the big bit first. The distance to the sun is totally outside the typical scale experiences of human beings. This doesn't necessarily mean we fail to understand it; it just means we need to take a moment to appreciate what this means for the size of our universe and our place in it. We can use analogies to help understand it, like how long it would take to drive or fly there, but the most useful way to realize the distance to the sun isn't an analogy at all. It's a fact. It takes *light* eight minutes and twenty seconds to travel from the sun to Earth. Imagine all the things that we could get done just in the time it takes a packet of light to burst out from the edge of the sun, hurtle its way through the inner solar system, and then bounce its way through Earth's atmosphere and into our eyes.

The Distance to the Nearest Stars

The scale of our solar system rapidly fades into insignificance when we consider the importance of the astronomical unit for estimating other distances. We know Earth's distance from the sun and that Earth orbits the sun, so we can simply and carefully watch the positions of all the stars throughout the year for the parallax effect. As the year progresses, the orbit of Earth takes us from one observing location within our solar system to another. Every six months, we are at a position two astronomical units away from where we started—on the other side of the mild ellipse that traces out Earth's orbit around the sun. This means that if a star is close enough, it will show the parallax effect, and that over the course of a year, nearby stars will appear to wobble back and forth on the sky with respect to stars a lot farther away.

It turns out that, compared to one astronomical unit, the other stars are massively far away. So very far away that very few show any significant parallax. Let's consider two examples: the brightest star and the closest star (they are not the same thing). Sirius A is the brightest star in the night sky. It has a parallax of 0.4 arcseconds (0.0001 degrees). The nearest star, on the other hand, can't be seen with the naked eye, because it is small, cool, and faint. It is called Proxima Centauri. It sounds sort of exotic, but it is merely the nearest star in the constellation Centaurus (the one that is supposed to look like a half-man, half-horse). Proxima Centauri has a parallax of 0.8 arcseconds, so it is *closer* than Sirius A.

The parallax angle is used in the same $\theta = s/r$ trigonometric relation we have seen many times before. In this case, s can be the one astronomical unit baseline and r the distance to a nearby star. If a star shows a parallax of one arcsecond when we observe it from a baseline of one astronomical unit, we find the distance to that star by dividing the baseline (one astronomical unit) by the angle (one arcsecond expressed in radians). There are 206,265 arcseconds in a radian, and therefore, there are 206,265 astronomical units of distance to an object that has a parallax of one arcsecond. We shorten the description of this parallax arcsecond distance to *parsec*, and it is equal to 206,265 times 150 million kilometers. One parsec is a staggering thirty-one trillion kilometers.

In determining the distance to the nearest stars, the familiar angle relation has been modified to $r = 1/\theta$. When we calculate the parallax of an object in arcseconds, its distance in parsecs is simply the inverse of the parallax, i.e., *r (distance in parsecs) = 1/p (parallax angle in arcseconds)*. If a star has a parallax of 0.1 arcseconds, it will be at a distance of ten parsecs (1 divided by 0.1). Proxima Centauri is at a distance of $1/0.8 = 1.3$ parsecs, and Sirius A is at a distance of $1/0.4 = 2.5$ parsecs.

Suddenly, this makes the distance to the sun look insignificant. It may take light eight minutes and twenty seconds to travel between Earth and the sun, but it will take light over four *years* to travel one parsec. It was difficult to make an analogy for the "short" distance to the sun, because that distance is orders of magnitude outside the experience of humans. However, we've now just upped the ante again by another six orders of magnitude simply by finding the distance to the next nearest star. Compared to the rest of space, however, that's just peanuts.

Astronomy Saves the World

Chapter 6: Space and Time

Human civilization is about ten thousand years old[1]. The human species is perhaps about two hundred thousand years old[2]. The oldest pre-*Homo sapiens'* tools found so far are about three million years old[3]. The dinosaurs went extinct about sixty-six million years ago. America and Africa were part of the same continent about two hundred million years ago[4]. Earth's atmosphere became oxygenated about two and one-half *billion* years ago[5]. Life on Earth began over three and one-half billion years ago[6]. Earth and the sun formed together about four and one-half billion years ago. Our galaxy is about thirteen billion years old, and our universe is about fourteen billion years old.

Compared to typical astronomical timescales, we are not old, and never will be. Use your time wisely.

Our atoms, on the other hand, are old. They are predominantly hydrogen, carbon, nitrogen, and oxygen. The hydrogen, and most of the helium, in our universe was created shortly after it all began. The carbon, nitrogen, oxygen, and other heavy elements were likely created around the same time Earth and the sun formed.

How do we know our universe is as old as it is? To answer that question, we need to do some dimensional analysis (i.e., consider some units), we need to know how to measure large distances, and we need to know how to determine the speed at which other galaxies are moving away from ours. An alternative approach is to simply look for the oldest thing in our universe, because naturally, our universe would have to be older than something found in it. However, this could only ever provide a lower limit to the age of our universe. Plus, as we shall discuss, it is rather difficult to calculate the ages of the stars and galaxies that we find.

How do we know the age of Earth? In this case, we are able to test parts of Earth using radiometric dating, which is another piece of nuclear physics. Before that, however, we need some more background on the distances we can measure. So far, we have only just scratched the surface.

Stepping up the Distance Scale

Until now, our distance estimates have relied upon only the understanding of angles as determined by simple triangles and circles. We have seen the parallax effect from observing the apparent shift in the position of an object on the sky as Earth orbits around the sun. However, as with most things in science, the effectiveness of this method is limited by the precision of the instruments used. So, exactly how accurately can we determine the positions of things on the sky in order to measure their distance using Earth's orbital parallax?

When viewed from the ground, the stars twinkle, because their light must pass through Earth's atmosphere. The twinkling is caused by their change in apparent brightness and apparent position as the rays of light get distorted passing through turbulent pockets of warmer and cooler air. The effect is called *refraction*, and the same thing is seen when we look at an object through water. For stars, atmospheric refraction means that if we want to find very precise positions, we must either find a way to get around the twinkling or make a correction for it. One obvious way around this problem is to put a telescope in space.

In 1989, about eight months before the deployment of the *Hubble Space Telescope*, ESA launched the High Precision Parallax

Collecting Satellite (*Hipparcos*[7]). The similarity of this satellite's name with Hipparchus of Nicaea is not a coincidence. He is the Greek astronomer who not only calculated the early trigonometric tables but also formulated a star catalog and stellar magnitude system. The spacecraft's four-year mission allowed us to catalog the positions of about 120,000 stars to a precision of 0.001 arcseconds (one milli-arcsecond).

Basically, the *Hipparcos* telescope looked in two directions at once while slowly spinning. While this type of behavior in humans might be considered amusing, it can be used as an advantage when determining the precise positions of stars. Instead of directly staring at a single star, the photons from two stars that are spaced far apart on the sky come together at the telescope's focus. At this point, a grid is put in the way before the light gets to a camera. As the spacecraft rotates, the starlight blinks on and off as the stars move across the grid. Slight differences in the relative positions of each star make correspondingly slight differences in the way the stars blink behind the grid. If the blinks from two stars were synchronized during one observation, but were not synchronized in an observation some time later, the position of one of the stars must have changed. As a result, the precision of the stellar positions is limited by how well we can monitor the blinking light, not the physical size of the telescope.

The 0.001 arcseconds precision data from *Hipparcos* means we have parallax measurements on stars with distances of about one thousand parsecs ($r = 1/\theta$.). To put that into perspective, it is about eight thousand parsecs to the center of the Milky Way. Andromeda, our nearest galactic neighbor, is at a distance of about eight hundred thousand parsecs.

In the grand scheme of things, 120,000 stars isn't that many, but they are enough to be useful. Within the *Hipparcos* catalog, we find many useful calibration stars that can be used to take another step up our distance ladder. One way to do this is to find out how bright something really is and compare it to how bright it actually looks.

Importance of Brightness

Distance and apparent brightness are inextricably linked. A candle one meter in front of us is quite bright. However, place it one hundred meters away, and it will only be a faint smudge of light. The candle hasn't actually changed its absolute intrinsic brightness; it just appears that way because of the different distances from which we look at it. The same considerations must be made in astronomy. The sun, which is one astronomical unit away from Earth, is almost overwhelmingly bright. If the sun were *only* ten parsecs away (about two million astronomical units), it would be a faint point of light unremarkable against the background of the thousands of other stars visible with the naked eye.

In the case of a candle, the light source is obvious. When looking at stars, this is not necessarily the case. Two completely different stars at two completely different distances can appear to have the same levels of brightness. Recall the example of the stars in Orion's Belt: Alnitak, Alnilam, and Mintaka? Alnilam is twice the distance of the other two, but they all look like the same dots of whitish light.

There are distances at which some stars are so far away that the light coming from them will be too faint to detect. It is not that these stars don't exist; it is simply that they cannot be seen. This means there is a fundamental bias in astronomy about what can and cannot be seen and, therefore, what we can and cannot easily study. It is not that our universe is made up of only bright things that are far away; it is simply that we are biased toward only seeing those things. We call this the *Malmquist* bias, after the astronomer Gunnar Malmquist (1893–1982) who first realized that this is a huge problem. Biases like this make astronomers very wary of data. We always have to ask, "Are we seeing everything that is out there?" or, "Do we have *all* the data?" The answer to both questions is usually an emphatic no, and that's okay, because we can deal with it. Our understanding of the limits of the data guides our interpretation of what the data is showing us.

How the apparent brightness of something changes with distance can easily be understood by considering the sphere. Once light emerges from a star, it will travel out equally in all directions. Each beam of light will travel along the radius of a sphere whose center is the star from which the light escaped. As two adjacent beams of light get farther and farther away from the star, their distance from one another also increases. We can see the same effect by simply holding up our index and middle fingers in a V shape. While waving our fingers around in this manner can either signal peace, victory, or our delight at helping Henry the fifth defeat the French army at Agincourt, it is also a useful demonstration about how the distance between two lines emerging from the same point will diverge if they are traveling along two radii[8].

The apparent brightness of something can essentially be thought of as the number of light beams reaching our eyes. More light beams equal a brighter light. As we get farther from the light source, the number of light beams falling on us from that source decreases, and so the object appears less bright. Unlike the two-finger example, which only deals with two dimensions, the light beams are coming from the stars in three dimensions and are spreading out along an ever-increasingly large cone.

When all the light from the star is considered, the same amount of light is being spread out over the surface of a larger and larger imaginary sphere. As a result, the light is being distributed over the surface area of a sphere given by four times π times the distance from the center of the sphere (the radius) times itself (radius squared), or $4\pi r^2$. This means that if we double our distance from an object, its apparent brightness will drop by four times (two squared). We aptly call this the *inverse square law* (see Figure 6.1).

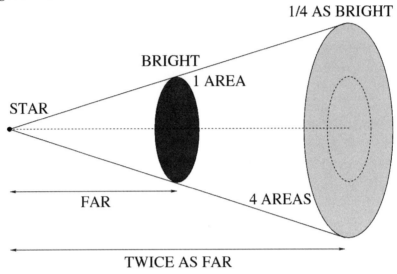

Figure 6.1: The inverse square law. Things that are twice as far away receive only a quarter of the light.

The inverse square law is remarkably useful when we want to estimate how far away something is, as long as we are confident about the absolute brightness of what we are looking at. If we know the absolute brightness of something and we measure its apparent brightness, we can easily calculate how far away that thing is.

Consider a simple lightbulb with a power of one hundred watts. The watt, named after James Watt (1736–1819), is the unit of power or *luminosity*. It is defined as the energy per unit time, or one watt is one joule per second. If we absolutely know that we are looking at a hundred-watt lightbulb, we can measure how bright it appears and use that measurement to determine how far away the bulb is. When we make our measurement, we only collect the energy per unit time coming from that lightbulb over a small area, i.e., a segment of the surface of a sphere. If we were to collect *all* the energy per unit time (one hundred watts) coming from the bulb, we'd need to use a measuring device that completely surrounded the bulb. Fortunately, the inverse square law means that we don't have to take such measures, and it also means that we can calculate how far away that bulb is as a result.

Until the invention of the telescope, the only area over which we could collect light was that of the human eye (about one square inch). This isn't a huge area, and one of the reasons humans can't see very well at night. It's also why most nocturnal animals have comparatively larger eyes than humans do. It's to help them spot predators and forage around for food in the middle of the night. While their eyes may also be adapted to using longer wavelengths than ours, the larger eyes do collect more beams of light, which in turn allows them to see fainter things.

The problems associated with observing faint sources are primarily why astronomers have a habit of building bigger and bigger telescopes. Telescopes are simply our way of giving ourselves really big eyes. Really big telescopes allow us to see intrinsically faint things that are close or intrinsically bright things that are really far away.

Once a telescope gathers the light, it is focused down onto a camera that is essentially a square of semiconducting material that tells a computer how many packets of light have hit it. That square of material has an area, so the camera is really telling the computer how many beams of light are hitting it per second per unit area. Therefore, we have yet another unit, which is called *flux*. This measures the energy per unit time per unit area. This flux is therefore equal to the luminosity, L (power of the lightbulb), divided by the surface area of that imaginary sphere around the object at that distance, i.e., $L/4\pi r^2$.

If we were to measure the light from our hundred-watt lightbulb at a distance of ten meters, we'd collect $100/4\pi(10^2) = 1/4\pi$ watts per square meter. If we knew we were observing a hundred-watt lightbulb and *measured* $1/4\pi$ watts per square meter, we'd know we were ten meters away from the bulb. The question then becomes, "Are there any objects equivalent to a hundred-watt lightbulb out there in our universe?"

It is here that we return to the 120,000 or so stars in the *Hipparcos* catalog. Each one of them has a parallactic distance measurement as a result of the apparent wobble in their positions. Since we know the size of an astronomical unit, we know the distance to these stars and can easily calculate whether they are the equivalent of ten-watt bulbs, hundred-watt bulbs, or thousand-watt bulbs.

We actually describe the luminosity of stars as compared to the sun. One reason is that the sun puts out about 380 million billion billion watts, so saying "one solar luminosity" is simply easier[9]. Earth is one astronomical unit from the sun, so at the top of the atmosphere, it receives a flux of $L/4\pi(1\ a.u.^2)$. This is about fourteen hundred watts per square meter. This is why solar power is a very viable source of renewable energy, which is something that should be more heavily invested in if we are going to begin to change our ways and move toward a more sustainable future for the coming generations.

Knowing the luminosity of the sun means that if we simply look for other sun-like stars, we will know their luminosities. We can then use the inverse square law to calculate their distances from Earth. This approach has a few problems. The sun isn't that luminous when compared to the most luminous stars, and there aren't that many stars as luminous as the sun.

Within five parsecs of the sun, there are seventy-four known stars. Of these, only *two*, Alpha Centauri A and Tau Ceti, are even remotely similar to the sun. Only *three*, Alpha Centauri A, Sirius A, and Procyon, have luminosities greater than that of the sun. Sirius, for example, is the brightest at twenty-four times more luminous than the sun.

In the entire *Hipparcos* catalog, about 50 percent of the stars are cooler and fainter than the sun. But, we must remember to be careful with statistics in astronomy, especially when they potentially involve the Malmquist bias. Remember, it is difficult to see faint things that are far away.

Naturally, the *Hipparcos* catalog contains far more bright stars than faint ones. But, in reality, most stars are hundreds of times fainter than sun-like stars, and there are far more of them. Faint things, even when there are loads of them, aren't going to be a great tool for determining large distances. Faint candles are no good. We need bright *standard candles*.

Standard Candles

It turns out that there are indeed some objects akin to stellar candles, and they are called *Cepheid variables*. These stars are thousands of times more luminous than the sun, and they pulsate so that the light coming from them gets brighter and fainter in a regular fashion.

What does the name Cepheid mean, though? There is a long history behind the names given to stars, which stretches back to the ancient Greeks and in and around the city of Baghdad during the tenth century (a period known as the Islamic Golden Age[10]). Sirius, for example, is basically the ancient Greek term for "scorching." Rigel is the loose Arabic translation of "foot" (the foot of Orion), and Betelgeuse is supposedly a translation from Arabic for "the hand of Orion." However, a colleague and friend of mine from Baghdad once assured me that "armpit" is the modern meaning. Considering the location of Betelgeuse in the constellation Orion, armpit makes more sense.

Before long, finding meaningful names for the stars, especially the faint ones, became a bit of a challenge. In 1603, Johann Bayer (1572–1625) decided it was time someone systematically cataloged the visible stars and provided a relatively easy way of identifying them. His method was fairly simple: Use the name of whichever of the eighty-eight official constellations the stars fell into, and then rank the stars according to their relative apparent brightness using the Greek alphabet[11]. Alpha Orionis is therefore the brightest star in the constellation Orion (which happens to be Betelgeuse).

In 1784, a curious star was spotted in the constellation Cepheus, which is named after yet another ancient Grecian mythological creature. Delta Cephei, the fourth brightest star in the constellation, was observed for several months by John Goodricke (1764–1786), since he knew that the brightness of some stars varied. He noticed, however, that the brightness of Delta Cephei varied in a very specific way. It rose in brightness quickly and became fainter over a longer time period. It was (and still is) as if the star were actually breathing in and breathing out. A short sharp inhale followed by a long exhale. Before long, there were many more of these Delta Cephei types of variable stars, which became collectively known as *Cepheid variables*. In fact, Edward Pigott (1753–1825), a friend of John Goodricke, had also discovered one of these variable stars shortly before Goodricke's observations of Delta Cephei. Pigott had noticed rhythmic variations in the star Eta Aquilae.

So what is all the fuss over these Cepheid variables? How can a variable star be used to determine distance if their luminosities are not constant? How can they be standard candles? First, they are really bright. Thousands of times brighter than the sun. This means they can be seen at great distances—including in other galaxies. Second, the length of time it takes for them to get brighter and dimmer is directly related to their absolute luminosities.

How could something have an absolute luminosity if its luminosity is changing? In practice, this isn't really a problem, because we can choose how to make the calculation of brightness be consistent each time. Using the maximum, minimum, or the average brightness are all obvious choices, but the average is preferred, as it gives the most consistent results.

Despite the discovery of Delta Cephei's periodicity in 1785, it wasn't until 1908 that Henrietta Swan Leavitt (1868–1921) discovered the relationship between the periods of Cepheid variables and their average *apparent* luminosities[12]. Remember that for accurate *absolute* luminosities to be determined, excellent parallactic distance measurements are needed. To make things a little trickier, Cepheid variables are a comparatively rare type of star, and there aren't that many close enough to show appreciable parallax using small ground-based telescopes. Leavitt's findings, however, didn't rely on nearby stars; she used repeated photographic observations of nearly two thousand variable stars in the Magellanic Clouds. These are two small galaxies that are very close to, and orbiting, our own. They can be seen as two quite obvious "smudges" on the night sky of the southern hemisphere, and they really are quite stunning.

The importance of the Magellanic Clouds for Leavitt's work is that their diameters are insignificant when compared to how far away they are. This means that the stars contained within them are all basically at the same distance. So any relative differences in the apparent luminosities of the stars correspond to the actual differences in absolute luminosity. Leavitt carefully gathered the data on the Magellanic variable stars, and as many astronomers do, she proceeded to see if any of the data was related. When the periods of the variable stars were compared to the luminosity, a period–luminosity relation was found. To this day, Leavitt's findings are startling, profound, and incredibly important. Cepheid variables are standard candles, and her work forms the cornerstone of modern cosmology.

Due to the difficulties associated with determining absolute distances in astronomy, the Cepheid period–luminosity relation has only been accurately calibrated relatively recently. This has been a result of the rise in space-based astronomy in the late 1980s and early 1990s.

Edwin Hubble (1889–1953) was one of the first astronomers to capitalize on Leavitt's work when in 1923 he noticed Cepheid variables in a number of *spiral nebulae*, which some had thought were in our own galaxy[13]. Hubble's use of Leavitt's standard candle variable stars showed that these nebulae were far outside our galaxy. The Andromeda Nebula became the Andromeda Galaxy. It was when Hubble started looking at more and more of Leavitt's variable stars in more and more galaxies that things started to get really interesting.

Astronomical Rainbows

For Hubble to complement the variable star data needed to uncover one of the most startling facts about our universe, he needed to use an instrument that could examine the electromagnetic spectra of galaxies in detail.

Compared to the telescopes a few inches in diameter that Galileo was making in the early seventeenth century, Hubble was able to use a telescope one hundred inches in diameter in the early twentieth century (the Hooker telescope just outside of Los Angeles[14]). Not only do large telescopes provide much clearer images, but they also collect much more light. This allows us to either detect much fainter objects or break up light into its individual colors.

Our eyes are only capable of seeing a sliver of light from the whole range of the electromagnetic spectrum. Most objects in our universe produce their own spectra, and some reflect spectra from other objects. We say that each object has its own *spectral energy distribution*, which is the fancy way of saying how the brightness of something changes with wavelength. The spectral energy distribution of an object can, depending on how the spectrum is produced in the first place, start with very low-energy radio waves and go all the way to very high-energy gamma rays. If we were to consider the whole electromagnetic spectrum compressed into the range of human hearing, the radio waves would represent the low-pitched notes of a large pipe organ, and the gamma rays would represent the high-pitched squeaks of bats as they navigate at night.

The wavelength, frequency, and energy of light are all interchangeable, and one need only know two numbers to convert them: the speed of light and the *Planck* constant. Astronomers use these properties interchangeably and typically report the radio and millimeter parts of the electromagnetic spectrum in terms of frequency; the infrared, visible, and ultraviolet parts in terms of wavelength; and the X-ray and gamma ray parts in terms of energy.

When there is a rainbow on a rainy day, we can see a small part of the spectral energy distribution from the sun. An instrument that can make astronomical rainbows is called a *spectrograph*, and the data from these can tell us a lot about an object. This includes a star's surface temperature, the absolute and relative amounts of chemical elements present, and the pressures and densities. By using the Doppler effect, we can use a spectrograph to determine how fast objects are moving toward or away from us, how fast objects are spinning, and some information about magnetic fields. It sounds like this information might be quite difficult to get, but all we need to do is pay close attention to what are basically the rainbows from each object.

Christian Doppler (1803–1853), the man who discovered the Doppler effect, reasoned that when an object is moving toward an observer, any waves it emits bunch together. When an object is moving away from an observer, any waves it emits stretch out. The object is catching up with the waves emitted forward and racing away from the waves emitted backward. For sound, this means an observer hears bunched-up higher-pitched waves ahead of a source and stretched-out lower-pitched waves behind the source. The effect is very obvious when we listen to the sound of sirens as an emergency vehicle passes. The same is true for light. If an object is approaching, the observed wavelengths are shorter and bluer—the object has a *blueshift*. If an object is moving away, the observed wavelengths are longer and redder—the object has a *redshift*.

As we have seen, stars contain a cocktail of chemical elements as the result of thermonuclear fusion. Photons generated in the core, that have made their way through the outer regions of a star, must still pass through an atmosphere that contains this cocktail before reaching our spectrographs. This means that some photons will interact with these chemical elements by giving their electrons energy. When photons do this, they disappear from the spectral energy distribution. An electron attached to a calcium atom absorbs the energy that a particular photon had. As we know the atomic structure of calcium, and we can test it in the lab, we also know the rest wavelength these missing photons must have. As a consequence, when we look at stellar spectra, we can identify dark bands corresponding to particular electron energy levels of known elements at known wavelengths. Joseph Fraunhofer (1787–1826) started to catalog these dark bands in the sun's spectrum. As a result, these stellar absorption lines are known as *Fraunhofer* lines, and they provide a mechanism for gaining a deep understanding of stars and galaxies.

If a star is moving toward or away from us, then the observed wavelengths of the absorption lines in their spectra shift because of the Doppler effect. If we know the difference between the observed wavelength and the known laboratory rest wavelength, we can easily calculate the velocities of objects as they move. The velocity is equal to the relative change in wavelength (the difference between the observed and rest wavelengths, divided by the rest wavelength) times the speed of light. The relative change in wavelength also allows us to quantify redshift. If the observed wavelength is 10 percent redder than the rest wavelength, the object has a redshift of $z = 0.1$.

Galaxies, those blobs of dark matter gravitationally gluing together billions of stars, have spectra that are the combination of billions of individual stellar spectra. That combination includes all the stellar absorption lines as well as lines that are smeared out from all the individual Doppler shifts. A typical galaxy spectrum is a whole mess of stellar spectra blended together. These spectra provide the relative velocity of the galaxies themselves, and it is this data that Hubble was able to combine with the Cepheid variable observations. The data did not disappoint.

Our Expanding Universe

When Hubble looked at the distances to the galaxies derived from Cepheid variables and the velocities of the galaxies derived from the Doppler effect, he found that almost all galaxies are moving away from us[15]. What's more, galaxies that are farther away are moving away faster. The relationship between the relative velocity of a galaxy and its distance is known as *Hubble's law*, and it plays a fundamental role in understanding that *our universe is expanding*.

We must be careful when interpreting the observations that show almost every galaxy moving away from us, though. It does not mean we are at the center of our universe. The key is to remember that farther away galaxies are moving away faster. A galaxy at a distance of one hundred megaparsecs (one hundred million parsecs) is moving away from us at seven hundred kilometers per second (a redshift of $z = 0.023$). A galaxy at a distance of two hundred megaparsecs is moving away from us at fourteen hundred kilometers per second (a redshift of $z = 0.047$). There are several analogies that can be used to describe this effect: dots on a rubber band; dots on an expanding balloon; and raisins rising in a muffin, cake, or loaf of bread as it's being baked. The rubber band analogy is probably the easiest to test. Draw five dots on a relaxed rubber band. Keep your eye on the center dot while stretching the band. All dots move away from the center dot with the outside dots moving farthest. Now keep your eye on a dot to the left or right of the center dot, and stretch the band again. All dots move away from that dot with the farthest dots moving the farthest. If there were a very long line of dots, the same conditions apply to all dots. No one dot is special. The same is true for galaxies. If some distant alien race in a different galaxy makes the same observations of the distances to other galaxies and their velocities, they would see exactly what we do. They could wrongly assume that their galaxy is the center of our shared universe.

While the rubber band, balloon, and cake analogies do not reproduce the observed velocities that the galaxies have, they do demonstrate that the galaxy observations are caused by the expansion of our universe. This means that our universe was smaller in the past. If we can measure the rate at which it is expanding, we can calculate how long it has been since all galaxies were in the same place at the same time, and we can calculate the age of the universe.

The actual data used by Hubble is shown in Figure 6.2. It may not look like much, but there is a noticeable lack of black spots in the top left and bottom right corners of the diagram. In this case, the top left of the plot would show galaxies that are nearby and have a large velocity, and the bottom right would show galaxies that are far away that don't have a large velocity. There are none. Our galaxy resides in a fairly lonely part of our universe, so we know that there are no nearby galaxies with large velocities, because we are capable of detecting them. In addition, many studies since Hubble's 1929 paper have surveyed many millions of galaxies, and not one of them has a large distance without a large velocity.

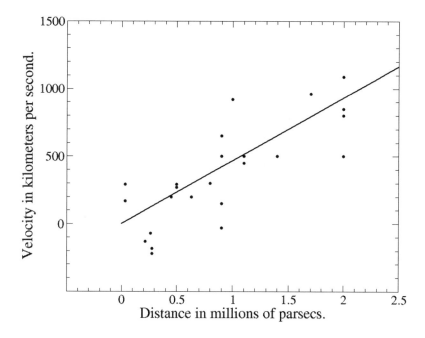

Figure 6.2: Reproduction of Hubble's law using the original, and flawed, data. The black line shows the trend in the data and also determines Hubble's constant.

Some of the galaxies in Figure 6.2 create vertical lines. These are galaxies within a cluster of galaxies, meaning that they all have the same distance but are moving around within the cluster so that a range of velocities is recorded. We can also see that some of the galaxies have negative velocities. That is okay, too. It just means they are very close to us and are actually moving toward us. Our nearest major galactic neighbor, the Andromeda Galaxy, is the lowest velocity point. We are moving toward each other at three hundred kilometers per second. That means it will be about four billion years before these galaxies collide, but by that time, Earth will probably no longer be habitable. The two points with the closest distances, above the solid black line at the far left, are the Small and Large Magellanic Clouds from which the Cepheid variable period–luminosity relation was first determined by Leavitt.

The distances used in Figure 6.2 are very inaccurate. As an example, look at the four points in the vertical line all the way to the right. These are all galaxies in the Virgo cluster of galaxies. That galaxy cluster is now known to be at a distance of 16.5 megaparsecs, rather than the two megaparsecs reported by Hubble. This is a huge, but explicable, discrepancy.

We shall return to the problems with Hubble's distances in a moment, but first, we will concentrate on that thick black line in the diagram. If all of the twenty-four data points in the plot could be squashed down fairly into one line, that thick black line would be it. It represents the "best fit" to the data. If we step up in the plot, this line tells us by how much we should step to the right. If our up step is 930 kilometers per second, then our right step needs to be two megaparsecs. Therefore, the slope of the line is 930 kilometers per second divided by 2 megaparsecs. This gives us 465 kilometers per second per megaparsec. This slope is known as *Hubble's constant*, and it is all we need to know in order to calculate the age of our universe. However, 465 kilometers per second per megaparsec is not at all accurate.

Age of Our Universe

Once we have Hubble's constant, all we need to do to calculate the age of our universe is some dimensional analysis. In this case, we have two units of distance and one of time. The analysis is simply the act of removing the redundancy in the two units of distance. The units of Hubble's constant are kilometers (a distance) per second (a time) per megaparsec (another distance). There are two units of distance that are divided by each other. The two distances are redundant and cancel each other out. This means that Hubble's constant is in units of inverse time, and all we have to do is invert Hubble's constant to get the time since all galaxies were in the same place. To do this, we need to divide the number of kilometers in a megaparsec by Hubble's constant, and out pops the age of the universe in units of seconds: 3.086 x 10^{19} divided by 465 equals 2.1 billion years. This answer is wrong, because we now know the age of the universe is about fourteen billion years, but the dimensional analysis is done and can be redone for any new estimates of Hubble's constant.

The biggest reason for the error in the first estimate of Hubble's constant stemmed from the difficulties involved in measuring large distances. There are only a few Cepheid variables close enough to the sun to get accurate distances from ground-based telescopes using parallax. Delta Cephei itself is at a distance of about 270 parsecs with 0.004 arcseconds of parallax.

Another issue fogging the calibration of the period–luminosity relation is all the intervening gas and dust along the way to any Cepheid. The gas and dust between the stars, the interstellar medium, actually makes objects appear fainter and redder. So, even though the *Hipparcos* satellite gathered the most accurate distances to a large population of bright stars and the *Hubble Space Telescope* picked out individual Cepheid variables in other galaxies, a lot of legwork was needed before we could settle on an accurate value of Hubble's constant. We now know it is seventy kilometers per second per megaparsec. Thus, 3.086 x 10^{19} divided by 70 equals 14 billion years: the age of our universe.

One of the interesting things about dimensional analysis is that we can ask some what-if questions. What if this number is actually zero? What if this number is really big? Does my model still make sense? In the case of Hubble's constant, we can use its dimensions not only to ask what if but also to devise a demonstrable sense of scale to our universe.

Recall that Hubble's constant has dimensions of length per second per length. One length describes a velocity. One length describes a distance. The units are, however, kilometers and megaparsecs. We can easily convert these to meters, or millimeters, and convert the seconds to years.

Astronomy Saves the World

Take two objects (hands will do) and hold them up about one meter apart. This gap represents the megaparsec unit. Due to the expansion of our universe, those hands would move apart by one millimeter over the course of about ten million years. However, hands have mass, and they mutually attract each other by their own gravity. If they started from rest and had masses of about a half kilogram each, they would collide in the middle after just two days. So, this explains why Andromeda and the Milky Way are moving toward each other. They are close enough for their mutual gravitational attraction to overcome the expansion of our universe. However, two galaxies that are much, much farther apart are going to be moving away from each other much, much more quickly. In this case, the expansion wins, not gravity.

The fact that our universe is not infinite and unchanging has caused some significant backlash from those groups that hold contradictory worldviews. This is incomprehensible, since the expansion of our universe is an observation that makes a prediction we can test. It is also backed up by an independent observation that makes the same prediction.

This second observation is the relative amount of helium compared to hydrogen. Hydrogen is one proton and is the simplest possible element. We know that helium is formed at the cores of stars as they fuse hydrogen, but there is far too much helium observed in our universe to be explained by stellar evolution alone. To account for the overabundance of helium, our universe must have been very hot and compact in the past so that helium could have fused from hydrogen all over the place. This hot, dense prediction is the same as that made by Hubble's expansion. Something that is cooling and expanding must have been hotter and smaller in the past. This hot, dense past is what we mean when we use the term "big bang." The Big Bang itself must have left a direct signature. If our universe had a hot, dense past, then we can predict how we would be able to detect a time in our universe's history when it had cooled enough for light to pass through it. This prediction is matched *exactly* by the observations of the cosmic microwave background—an incredibly precise signal that comes from every direction at the precise wavelength predicted.

The evidence for the Big Bang is a perfect example of the scientific method applied. We have two independent pieces of evidence making the same prediction. When we search for the signal of that prediction, we find exactly what we are looking for. It is an open and shut case. As for a first cause for our universe, well, we are still figuring that out.

The observations about the age of our universe are, on paper, relatively simple. Find and properly calibrate enough standard candles in or around our galaxy. Find those standard candles in other galaxies to get their absolute distances. Get spectra of the galaxies to find their relative velocities. Plot distance versus velocity. Show our universe is expanding and was therefore smaller in the past. Fit a line, and use the slope of the line to find the age of our universe, and there we have it. The ages of other things, like galaxies and stars, aren't as straightforward.

There is no direct method for determining the age of our galaxy. Unlike our universe, there are no handy galaxy relations that can be rewound until a zero point is reached. We can try to estimate how long it may have taken multiple generations of massive stars to process the amounts of heavy elements we see in our galaxy. We can try to make a simulation of how long it may take a primordial gas cloud to collapse, form stars, form more stars, and then end up with a model galaxy looking like what we think our actual galaxy looks like. Or we can look for the oldest thing we can find and at least know that our galaxy must be older than that.

It is clear that our galaxy has undergone some merging, and that has caused some extra star formation and disruptions in how things move. However, the *globular clusters* surrounding and orbiting our galaxy have been mostly isolated from all this. These tightly packed, largely spherical systems can contain up to ten million stars that have mostly formed together and lived out their lives undisrupted. This means that the stars in globular clusters form very well-defined main sequences.

If we make the fairly safe assumption that the globular clusters formed at the same time as our galaxy, then we can look to see how many stars in a cluster are left on the main sequence. Remember that massive, bright stars live shorter lives than small, cool stars. If we can work out how long a particular mass star remains on the main sequence, where it is fusing hydrogen in its core, we can age the cluster. A young cluster will have all stars still on the main sequence. An old cluster will only have cool red stars left on the main sequence. The point at which cluster stars turn off the main sequence indicates the age of the cluster. Applying this technique to the 150 or so known globular clusters around our galaxy, we find the oldest has an age of about thirteen billion years[16]. This implies that our galaxy must have formed around the same time.

Age of Earth

As with globular clusters around the Milky Way, the sun, Earth, and all the planets in our solar system must have formed at the same point. They are essentially the same age. We know this, because we can directly see stars and planets forming together in other star systems. Therefore, if we can get a good age for one, we have a good age for them all. Determining the age of the sun directly, like determining the age of our galaxy, is very difficult. It depends on how well we understand the conditions of the sun and how good our models of the sun are. Even if these models are very accurate, we basically need to wait for the sun to evolve to test them, and that only happens over very long time periods. The sun is predicted to increase its effective temperature by about ten degrees over the next *billion* years[17]. Or, one degree over the next one hundred million years, and 0.01 degrees over the next million years. This is one of the reasons the sun getting warmer is not a viable explanation for the current global climate change trends.

Finding the age of Earth can be done by testing bits of it. We can look at how rocks have been layered from the settling of particles in long-lost oceans. We can clearly see a progression of older and older fossils back to less complex life forms embedded in deeper and deeper layers. And we can measure the continental drift and rewind the clock until we are left with one massive expanse of dry land—Pangaea. This single land mass existed around two hundred million years ago, and there are thought to have been even older supercontinents like the two-billion-year-old Colombia. Regardless, we can never be sure that we've found the absolute oldest layer of rock. So, while we can place some age limits based on estimates from Earth rocks, it'd be great to have another choice.

Our ability to determine the age of our galaxy by using global clusters has another similarity with how we can determine the age of our solar system. Globular clusters formed with the galaxy and went about their business without being influenced by all the stars forming, stars exploding, more stars forming, galaxies colliding, or general mayhem. The equivalent objects in the case of aging the solar system are asteroids. They were formed around the same time as the solar system, passed their time independently from any surface processes on Earth that might complicate determining ages, and are something that we can directly test in a lab when we pick them up as meteorites.

In Chapter 3, we discussed the thermonuclear processes that take place at the centers of stars. Low-mass stars slowly fuse hydrogen into helium, solar-mass stars mostly fuse hydrogen but get hot enough to fuse helium into carbon for a short while, and high-mass stars fuse everything into nickel, which then decays into cobalt and iron. Even more massive elements must be produced in supernovae. These catastrophic events also have the effect of distributing all these materials back into space, which then go on to form new solar systems. Most of our solar system's material went into the sun itself, and some went into the planets. The bits left over clumped together to make the asteroids, and some of those asteroids fragmented and bumped into one another to produce smaller meteoroids.

Given the supernova origin for the heavy elements in our solar system, it should come as no great surprise to find that all meteorites contain some amounts of nickel and iron. Most commonly, these two metals are mixed with large numbers of silicon chondrules to make a meteorite known as a *chondrite*. So how can we determine the age of a meteorite, or any rock for that matter? Most people automatically think of carbon dating, but that isn't technically correct. We use *radiometric* dating. For pinning an age on our planet, carbon isn't used at all.

We must now turn our attention back to nuclear physics. When massive stars burn their final nuggets of fuel, they have reached the point of producing, among other things, nickel-56. This decays to cobalt-56, and that in turn decays to stable iron-56. But this doesn't happen all at once. Just because a lower energy condition is available doesn't mean that an isotope will automatically find it. This decay will happen randomly and on a varying timescale.

If we were to take a single nickel-56 atom and stare at it, we wouldn't be able to predict when a proton was going to emit a positron for cobalt-56 to form. It could happen as soon as we look at it, or it could happen thousands of years from now. However, similar to observing billions of galaxies to watch stars explode daily, if we stared at billions of nickel-56 atoms, we'd see these decays taking place regularly.

As soon as we start talking about large numbers of anything, we quickly get involved with statistics. Like astronomical distances and astronomical times, statistics are very often misused or completely ignored. This can be by accident, on purpose, or through simple ignorance. Astronomers must deal with statistics regularly, because we do a lot of counting. How many stars are in that cluster? How many stars are in that galaxy? How many photons hit my camera? In many cases where counting is necessary, we don't count them all; there's just too many of them. Instead, we use the power of statistics to make things easier.

In the case of radioactive decay, we treat a mass of material in a statistical way. We don't count individual decay events. We look at the system as a whole. This way, we can ask how long it takes for half the nickel-56 isotopes to become cobalt-56. The answer is a little over six days. Notice that we don't need to have a specified amount of nickel-56 to start with; all we need to know is that it takes a little over six days for half the individual isotopes present to decay. The amount of time needed for a lump of unstable material to convert half itself to a lower energy state is called the *half-life*.

The half-lives of unstable isotopes give a method for aging old things. If we found the ratio of nickel-56 to cobalt-56 to be 50/50 that lump of metal would be about six days old. If we found the ratio of nickel-56 to cobalt-56 to be 25/75, the thing would be twelve days old. There is a useful analogy with the melting of water ice that can be easily demonstrated with some fairly simple equipment.

Take two very sensitive weighing scales and raise one a little higher than the other. Place an ice cube with a dash of food coloring in it onto a dish that will allow the melted water to run off onto another dish on the lower scale. As the ice melts, the weight on one scale gets transferred to the other scale. The ice "decays" into water. In this case, the decay from ice to water is a straight line. As the mass of ice goes down, the mass of water goes up, but the ratio of the ice to water tells us how long it has been since the ice started to melt. Some example videos of this process can be found online [18,19]. In the case of isotopes decaying, the ratio relationship is not a straight line. It is half and half and half again. This is an exponential decay, but the ratio of the isotope masses still provides us with an age.

To measure the relative number of elements in a lump of material, we need a *mass spectrometer*. This device takes the lump and removes electrons so that the excess number of protons gives the elements a net positive charge. Different elements will have different net charges. These elements are then fired into a magnetic field that deflects the paths of the elements in much the same way that a magnet deflects the needle direction on a compass. However, if there is a large net positive charge, the path deflection is large. If there is a small net positive charge, the path deflection is not as large. This means that two (or more) elements can be separated out and counted by charge-sensitive detectors. This provides the mass ratio of two or more elements, and therefore a way to age the lump.

As we already saw, the half-life of nickel-56 is six days, but the half-life of cobalt-56 is over seventy-seven days. This is nothing compared to some other elements. The uranium-238 isotope is unstable with a half-life of about 4.47 *billion* years. This makes it very useful for determining the age of really old things, like our solar system.

In 2010, Audrey Bouvier and Meenakshi Wadhwa carried out an analysis of a 1.5-kilogram chondritic meteorite found sometime during 2004 in northwest Africa. They used a mass spectrometer to measure the relative amounts of various isotopes including uranium-238 and lead-206. Uranium-238 actually follows a decay chain through isotopes of protactinium, radium, bismuth, and thallium to name a few. It finally settles at stable lead-206 after taking 4.47 billion years to complete the entire process. When Bouvier and Wadhwa measured the ratio of uranium-238 to lead-206, it showed the meteorite was 4.5682 billion years old [20]. Notice the five significant figures. This same age was confirmed using other decay chains, and so it places a tight constraint on the age of the solar system, the sun, and Earth of 4.5 billion years.

If someone contests the actual age of Earth because they say that carbon dating is unreliable, it is because they have been misinformed on the method of radiometric dating. The age of Earth isn't based on carbon dating—it is based on uranium–lead dating.

When dealing with an answer of 4.5 billion years, an uncertainty of a million years doesn't make a blind bit of difference. However, when dealing with tens of millions of years, a million-year uncertainty would limit how well we determine the absolute age of something. So, different radiometric dating isotopes must be used for different time periods based on their half-lives. Argon–argon is about 1.25 billion years, iodine–xenon is about sixteen million years, and carbon dating is, ironically, about six thousand years.

When we consider how accurately we can measure things, we must be careful with these isotope ratios. It is relatively easy to measure an isotopic ratio of 50/50, but as the material gets older and older, the ratios get more difficult to determine with high precision. A 50/50 measurement is a lot less tricky than a 1/5000 measurement. This is why different decay chains are used for different ages. And so, just as the staggering distances we find in our universe frequently leave us stunned, radiometric decay chains indicate timescales that have the same effect. We are able to confidently age objects in our universe, including Earth itself, that are monumentally old.

Astronomy Saves the World

Chapter 7: Fair and Balanced

Unlike the delightful contradiction of very rare and violent events continually happening throughout our universe (because of its massive size and age), in the vast majority of cases, astronomical objects exist in some sort of relatively simple balanced, or equilibrium, state. The realization of this common condition allows us to take the basic principles of force, energy, and momentum, combine them with our understanding of space and time, and use the scientific method to produce explanations for the astronomical objects we detect.

We can use ourselves as an example of something that is typically in physical equilibrium, but first, we must briefly revisit the subject of energy. When we are standing, the centers of our bodies are farther away from Earth than if we were sitting. When we are standing, we know it would be much more comfortable if we were sitting. It would be even more comfortable if we were lying down. It requires more energy to stand than to sit, and more energy to sit than to lie down.

If we were to fall over, our equilibrium would be lost. We would speed up as we fell and gain *kinetic* (motion) energy. The *potential* energy of gravity is exchanged for kinetic energy. While standing, we always have the potential to fall over, and our muscles must work to stop this potential energy from becoming kinetic energy. The muscles in the human body expend energy as they act against the gravitational force of Earth's mass pulling us toward its center. After standing for a while, we get tired and gravity starts to win. Gravity wins a lot in our universe, and in the absence of any external or balancing source of energy, a system will resort to the lowest possible energy state it can.

When something isn't speeding up or slowing down, getting hotter or colder, bigger or smaller, absorbing or emitting energy, that system is said to be in *equilibrium*. Two or more parts of that system have balanced so that on the whole nothing exciting happens, which is typically a great way to start solving a physics problem. Ask yourself, "Is the thing I'm looking at doing anything interesting?" If not, it is probably in equilibrium. If the system is doing something interesting, it is likely that one force is winning and making the thing accelerate, explode, collapse, or something else. That system is probably not in equilibrium, and an accurate description of what is happening could be a little trickier.

The forces acting on objects must all cancel if equilibrium is to be established. So, in the case of standing, the force of gravity is pulling us down, and the force of the ground on our feet is stopping us from falling into the surface of Earth. The muscles throughout our bodies are keeping our skeletons rigid enough, and our brains are telling them how to contract and relax in order to stay balanced. When our muscles are no longer able to keep our skeletons rigid or balanced, then the equilibrium condition is broken, and we fall to the ground. This is almost always interesting, sometimes funny, and sometimes tragic.

Our Universe

A falling human is between two equilibrium conditions: standing and *sprawled on the floor*. Is our universe also between two equilibrium conditions, or is it a fair and balanced system? We have seen the evidence for the Big Bang, we know our universe is expanding, and we have discussed how our universe is heading for a cold, sparse future. These do not seem like equilibrium conditions. We've already seen that our universe is under no obligation to be easy to understand, so unsurprisingly, the question about its potential equilibrium is not simple, and I offer no satisfactory answer. There are, however, some interesting talking points.

We must be mindful of asking nonsensical questions. What came before the Big Bang, and where did it happen? These questions don't really make much sense. They ask about an event in space and time before space and time existed and are the equivalent of asking, "What type of aircraft did dinosaurs use to go on vacation?"

We must also be mindful of the meaning of words we commonly use. "Infinity" and "nothing" are two excellent examples. By definition, infinity is the concept of a number that can never be reached. Nothing is also the concept of a condition that can never be reached. The nature of quantum mechanics essentially prohibits that idea from becoming a reality. If we take everything out of a volume of space, then there is still going to be *something* there. There is still energy in a vacuum. Part of the challenge in understanding these things is that the reality of "nothing" does not conform to our classical definition of "nothing." The concept of nothing is so misunderstood that our universe could have actually come from it[1].

If we avoid the nonsensical questions and carefully consider the words being used, we can begin to be somewhat constructive on the topic of our universe and equilibrium. The growth of our universe is causing the transition between the hot Big Bang and its frigid expansive future. We now know, however, that the expansion rate of our universe is not constant. Observations of distant exploding stars show that they are slightly fainter than they should be[2]. This means the expansion rate is increasing, and our universe is accelerating. The mechanism responsible for this feature of our universe has been dubbed *dark energy*. Something is stretching space itself. The jury is still out on what that something is, but this observation may in some way be linked to the vacuum energy present in supposedly empty space[3].

The energy in our universe gives us one last talking point. As nothing can enter or exit an isolated system, the total energy in our universe cannot change. So what is the total energy of our universe? By the observed motions of galaxies and stars and the light they emit, we might suppose that there is an awful lot of energy. But by adding in things like vacuum energies and dark energies, it could well be that the total energy in our universe adds up to zero. We would then be left with a case where our universe started from nothing, is nothing, and is going to become nothing. That seems at least somewhat fair and balanced to me.

Galaxies

The large-scale structure of our universe leaves quite a lot to think about, but for most other astronomical objects, the explanations we have are a little less unsatisfactory. Some of them still have their own mysteries, however, and galaxies offer a good example.

Are galaxies fair and balanced systems? As previously mentioned, galaxies come in two major categories: elliptical and spiral. Elliptical galaxies look boring, but spirals do not. Does this mean elliptical galaxies are in equilibrium and spiral galaxies are not?

Galaxies are collections of hundreds of billions of stars all being corralled around by gravity, and the distribution of their stars is complicated. The physical 3D shapes of galaxies determine the paths that groups of stars can follow as they orbit around their galaxy centers. Most stars follow different classes of orbit, and each class of orbit tends to have slightly preferential directions of motion. We can't actually *see* these orbits, but given some initial position, some initial motion, and an assumption of how the mass in the galaxy is dispersed, we can make computer models of galaxies and show what paths the stars can take.

The possible orbits of stars in galaxies are given names like boxy, peanut, and X-shaped, and they make quite beautiful 3D patterns. The orbits of these stars don't follow the typical closed orbits of planets, where one or more complete rotations later they are back in the same place. In fact, they will probably never return to the exact same place because of other events. Instead, the paths they follow are more like the complex patterns we make when using an intricate spirograph[4].

If these orbits weren't complex enough already, then there are two additional events that can happen to the stars and a galaxy itself to upset things even further. These are not single stars orbiting within an entirely smooth gravitational field intrinsically generated by a galaxy. Each star within the galaxy produces its own gravitational field, so the overall gravitational field is slightly bumpy on small (parsec) scales. This means that even though the stars are massively far apart compared to their diameters, they do gravitationally interact with one another and slightly deflect each other's paths. This means that instead of the orbits of stars being tied to their predefined boxy, peanut, and X-shaped paths, they deviate from these trajectories, and the orbits become more and more random over time.

After so many gravitational interactions, a star's motion will have been changed so much that any indication of its parent orbit will be lost. The length of time it takes to do this is known as the *relaxation timescale*, and for a typical star in a typical galaxy, it would take well over one hundred billion years to get to this point.

Another thing that may upset the orbits in galaxies is when they merge with other galaxies great and small. This happens to both elliptical and spiral galaxies. Minor mergers, between one large galaxy and one galaxy that is perhaps an order of magnitude smaller, can happen fairly frequently (every few hundred million years or so). In fact, our galaxy may still be adjusting itself after the last one. Major mergers between two comparable galaxies may happen every billion years or so. This is well before the relaxation timescale is reached. During a merger, the orbits of the stars are all shaken up, the galaxy that results from the merger has to settle down again, and the relaxation timescale clock is reset to zero. Note that, during a merger, the average distance between stars is still massive, and the chances of two stars colliding are still incredibly small.

The odd shapes of galaxies, the relaxation timescales, and the fact that they are being crashed into every once in a while make it almost impossible to imagine these systems ever reaching a typical equilibrium state. However, there is more to this story than literally meets the eye. Based on more detailed observations of galaxies, there are a number of resulting questions that can only be resolved using dark matter.

The details surrounding the existence of dark matter provide us with an example of how the nature of something can be determined without actually seeing it. In the case of dark matter, its presence is inferred through its gravitational influence. Now, we must be careful with approaches like this, because there are people in this world who will use inference, with no direct evidence, to make extraordinary claims about how their *opinion* must be the truth. This is absolutely not the case that supports the existence of dark matter.

Observations made by Vera Rubin during the 1970s showed that galaxies are rotating faster than expected purely from the mass of their stars[5,6,7]. With enough understanding of stellar astrophysics, we can comfortably say that one thousand sun-like stars will have the mass of approximately one thousand solar masses. But based on the mass assumed from the observed starlight alone, galaxies should be shredding themselves apart as they rotate. Instead, they are actually acting as if there is about six times more mass than can be seen in light. There must be unseen gravitational glue holding them together. Dark matter.

Some galaxies also seem to have odd 3D shapes. We can't see these shapes directly, as we can only observe how they look in 2D, but we can estimate them from the orbits that could reproduce the observed distribution of light. Shapes like prolate spheroids. Mergers are one reason they may have this shape, but they could also have an irregular distribution of dark matter.

In addition to observing the rotations and shapes of galaxies themselves, we also see that light from very distant galaxies gets warped as it travels through less distant galaxy clusters. This shows us that there is a distribution of unseen mass permeating the foreground clusters of galaxies. This *gravitational lensing* allows us to roughly estimate the distribution of this unseen mass and produce low-resolution maps of dark matter.

If that isn't enough, we also have grand computer models that can reproduce the evolution of our universe. Models with names like the Millennium Simulation Project, GADGET, and the Bolshoi Simulation[8,9,10]. Given the observed conditions present during the early universe, all these models require the presence of an unseen distribution of gravitationally interacting material in order to reproduce the same type of universe we see today. Dark matter.

These grand computer models have shown that dark matter must be distributed in mesmerizing filamentary patterns that have often been compared to the neuron patterns seen in the brains of mice. Just to be clear, despite threads fueled by anonymous amateur Internet philosophers, there are no relationships between the two. However, these computer-generated universes show that when two dark matter filaments join, there is a concentration of regular matter. This matter then forms into galaxy clusters.

Galaxy clusters contain about one hundred to over one thousand galaxies. Compared to the number of stars in galaxies themselves, this isn't many, and we can quite happily count all galaxies in a cluster individually. We can also measure how fast they are all moving around in the cluster itself and calculate the gravitational mass of the cluster. When we compare this to the amount of light coming from the clusters, we find that, like galaxies themselves, a massive amount of dark matter is required. This is the same dark matter that is required to reproduce the warping of light seen by gravitational lensing. While there may be the odd small galaxy that gets gravitationally slung-shot out of a galaxy cluster, for the most part, galaxy clusters are also being kept in some sort of balance by dark matter.

Similar to the independent observations we have to support the Big Bang—an expanding universe and a large amount of helium—dark matter is the simplest explanation for all these pieces of evidence. Should a better explanation come along in the future, and there have been some fairly unsuccessful attempts already, then the better explanation will eventually be adopted by the scientific community if the scientific method has been followed.

The mysterious nature of dark matter also provides another avenue of wonder for people interested in the sciences. This is an area rich with ongoing inquiry and offers the chance for many future PhD projects and a likely Nobel Prize.

Research into dark matter is bringing the forefronts of astrophysics, particle physics, and theoretical physics together, much like the study of our early universe. The favored explanation is that dark matter is made up of Weakly Interacting Massive Particles amusingly called *WIMPs*. The Alpha Magnetic Spectrometer experiment currently onboard the ISS is trying to find observational evidence for these dark matter particles. Some have suggested that a population of black holes left over from the Big Bang could produce the signature of dark matter [11]. Others have tried to modify our understanding of gravity when applied to very large scales [12]. The answer will come, eventually, but for the time being, we accept that dark matter is what keeps elliptical and spiral galaxies in some sort of balance, even if they are regularly disrupted by mergers.

The Motions of Stars

The vast majority of light in most galaxies comes from stars. These stars can range from hundreds of times larger than to less than one-tenth the mass of our sun. In most cases, they are relatively small, relatively cool, and relatively dim compared to our own sun. Most of these stars also have planetary systems, and it is likely that many of these planets are similar to Earth.

Barring any catastrophic astronomical incidents, the orbit of Earth around the sun is stable. In the same way that Earth's gravity pulls us toward its center, the sun's gravity is also pulling Earth toward its center. The same solar gravity that keeps Earth on its yearlong orbit also keeps the other planets, dwarf planets, meteoroids, asteroids, and comets on their orbits. But unlike an exhausted human falling to get closer to Earth, Earth is not getting closer to the sun. Something must be balancing its orbit. Quite simply, and perhaps a little counterintuitively, the gravitational pull of the sun is balanced by the motion of Earth itself.

Since we now know the length of the astronomical unit, it is quite easy to determine how fast Earth is actually moving around the sun. Ignoring the slightly elliptical nature of Earth's orbit, it is the circumference of a circle divided by the time it takes to complete that circle (speed equals distance divided by time). For Earth, this is two times π times an astronomical unit divided by the number of seconds in a year. Converted to an everyday unit, this works out to be about seventy thousand miles per hour. If we were to whirl a ball on a rope around our head, we would see the same principle at work. The ball is going fast, but the force of the string keeps the ball moving in a circle by pulling it toward the center. In the case of a planet on a stable, circular orbit, the force of the string is analogous to the constant pull of gravity from the star it is orbiting.

Much like how an unintentionally falling human usually has the misfortune of hitting the ground, humans aboard the ISS, and the space station itself, are also falling. They just happen to have enough horizontal speed to match how Earth's curvature drops away; they miss the ground. This does not mean that gravity has been turned off. The ISS orbits only a few hundred kilometers above the surface of Earth where the gravitational pull is still 90 percent of what it is on the ground.

One of the greatest scientific legacies of Johannes Kepler is his realization that the orbits of our solar system's planets are not circular but elliptical (Kepler's first law). But what rings true for circular orbits also applies to elliptical orbits. There is the same balance between gravitational potential energy and the motion of the object.

Not only are planets in elliptical orbits around stars, but stars can also be in elliptical orbits around one another. When we look at the stars, we see that many of them seem to come in pairs, triplets, or pairs of pairs, and the orbits of these binary and multiple star systems turn out to be incredibly useful.

A physical law we observe on Earth, or in our solar system, is a physical law throughout our universe. There has never been any evidence against this fact, and there is a mass of evidence to support it. Kepler's laws therefore also apply to binary stars and can be used to determine individual stellar masses. This is important, because as we discussed in Chapter 3, mass is the governing feature of how a star behaves.

Kepler showed that the period of a planet in years squared is equal to the distance of that planet from the sun in astronomical units cubed ($P^2 = a^3$). One can take this law a little further and add in the mass of the sun in solar units ($MP^2 = a^3$). This doesn't make a difference in our system, since multiplying by one solar mass doesn't change the law. Also, as Newton showed with his law of gravitation, including mass is necessary to get the dimensional analysis right. In other star systems, this additional mass becomes the total mass of the binary system in solar masses, and the distance between the two stars gives the ratio of the two masses.

With the total mass, and the mass ratio, calculating the individual masses of stars in binary systems can be trivial.

Hardly anything is ever that straightforward, though. The orbits of binary stars are going to be randomly distributed in 3D, and we will observe them with a 2D coordinate system projected onto the night sky. So, we need to know what the tilt or inclination of the system is in order to get an accurate calculation of their masses. We also need to know the distance between the stars in astronomical units. The absolute best systems for calculating individual stellar masses are the nearby binaries, in which two stars eclipse one another and can also be seen separately. By doing so, we would know the absolute distance using parallax, what the orientation of the orbits of these stars is along our line of sight, and what the distance between the stars is. There are not many of these systems, though, so we typically calculate the distance between the stars from spectra. The Doppler shift gives us information about how fast binary stars are orbiting one another and Kepler's third law is used again.

At least half the stars we observe are part of systems with multiple stars. Space is really big, so the average gap between stars is vastly different to the diameters of stars. The diameter of the sun is about ten million times smaller than the distance to the nearest star. So, just like the chances of two stars colliding, the chance of two independent stars randomly getting close enough to establish a stable binary is also very, very small. This observation tells us that when we see two or more stars orbiting one another, one did not gravitationally capture the other. They most likely formed like that. The in situ formation of binary stars implies that most stars form in groups.

The sheer number of stars in binary or multiple systems is not the only evidence we have that stars form in groups or clusters. We frequently find loosely clustered stars inside galaxies, and tightly clustered stars in discrete globs orbiting around galaxies. But the most spectacular direct evidence for stars being born in groups comes from observing possibly the most mesmerizing objects on offer—star-forming regions. These are clouds in space where we actually see stars forming together in groups. Images of these colorful, complex-looking, wispy nebulae are often used to decorate lecture rooms, classrooms, labs, and public places around the world. Famous examples include the Orion Nebula, the Eagle Nebula, and the Trifid Nebula.

The process of star formation starts when a vast cloud of interstellar gas and dust gets knocked slightly out of balance. Individual regions within the cloud then collapse until the resulting increase in pressure starts nuclear reactions to push back against the gravity. Equilibrium is restored when stars are born.

These clouds change from cold, dark blobs to wispy wonderlands as the first newly formed stars illuminate the remaining gas to create a star-forming nebula. As more young stars use up the gas, these objects become a collection of points shining brightly on the same patch of the sky. To witness the process over a few million years would be similar to bringing a blurred image of a forest into focus to see the individual trees.

The few thousand stars in these clusters all have their own motion within the group. Their random individual elliptical orbits within the cluster keep them from falling to the center of the cluster, but there is nothing physical for these stars to fall toward and collide with. The gravity of the cluster is created from all the individual stars within it, not a central object.

The stars in this type of open cluster would carry on their lives in orbits that keep the cluster balanced if it weren't for the fact that these open clusters form in rotating galaxies. This rotation provides an external influence on the cluster, making it so the balance between their motion and their mutual gravity can't be maintained. As the clusters get older, the stars within it get pulled farther and farther apart as it orbits around inside its galaxy. The spaces between the stars open up, and, like a drop of milk swirling around in a freshly brewed cup of Earl Grey tea, these open clusters dissolve into the general diffuse population of galactic stars.

Star clusters are also found outside galaxies where they do not get ripped apart so easily. These are the globular clusters used to constrain the ages of our galaxy in Chapter 6. They plod around the outskirts of galaxies, generally minding their own business. With these, the difference in the effect of the galaxy's own gravity, from one side of the globular cluster to the other, is not enough to disrupt the stars in the same way they are disrupted in open clusters. They are also balanced by a fair exchange between the total gravitational potential energy of all the individual stars combined and the average kinetic energy of all the individual stars.

The gravitational potential energy of a globular cluster is determined by the positions and masses of the stars. Their average motion determines the kinetic energy. As these are averages, the direction of motion isn't as important, because when the directions are averaged, they have, for the most part, an equally likely chance of moving with the same velocity in each of the three dimensions. Therefore, if we know the average velocity and the radius of a globular cluster, we can calculate its mass. The balance between these two energies occurs a lot in gravitationally bound systems, especially when they are spherical. This is called the *virial theorem*. The term likely comes from the Latin "vires," which can mean "forces."

We saw that the typical relaxation timescale for a typical star in a typical galaxy is over one hundred billion years. When the same relaxation calculations are done for typical globular clusters, we find a timescale of only a few hundred million years. Consequently, we can say that globular clusters are relaxed and balanced, thus justifying the use of the virial theorem. Typical galaxies are not relaxed. They can't be in a universe that is only about fourteen billion years old.

Remnants of Stars

By considering the balances involved, we can provoke an understanding of many astrophysical objects: clusters of galaxies, galaxies, globular clusters, open clusters. But as we saw in Chapter 3, stars themselves also exist in a state of equilibrium.

The equilibrium conditions that must be present in main-sequence stars have taught us a great deal about our origins and the fate of our sun. The history of the heavy elements, fused in the phenomenal conditions at the centers of stars much more massive than our sun, is to me remarkably profound. It also serves as a warning about our potential future. The elements that make us, and our planet, are here as a result of bloated stars greedily consuming the cosmic resources available to them as quickly as gravity would allow. Those stars are now dead. If they had been smaller, more modest, and less greedy, they would still be gently gracing the low-mass end of the main sequence. We owe our existence to the behavior of massive stars, but if we as a species mimic their example, we too could end up as a testament to the ignorant consumption of limited resources.

Not only does our understanding of stars provide us with another source of cosmic humility, our quest to determine how they work has pushed the field of nuclear physics forward. Julius Robert Oppenheimer (1904–1967), who directed the construction of the first atomic weapons, was guided by his findings and experiences studying post-main-sequence objects such as white dwarfs and neutron stars. While nuclear weapons have obvious and devastatingly evil applications, the same technology has gone on to provide non-fossil-fueled power-generation options. These are options that we will be forced to convert to over the coming decades if no significant measures are taken to harness truly green sources of energy like the sun. Once again, our understanding of astronomy has given the world a chance to avoid catastrophic climate change.

Despite what we do to our planet, our sun will evolve into a very different object in the future. A sun-like star may have spent ten billion years living in a core hydrogen-burning, main sequence, equilibrium state, and a few hundred million years puffing through the end of its thermonuclear life. However, it will spend way more than ten billion years in its final white-dwarf state.

Sun-like stars go through a fundamental change in their structure and their physical support mechanisms. Perhaps it is not fair to say that they die, and that the main-sequence and post-main-sequence phases are merely a warm-up to their mature selves. A rather lazy analogy of a child becoming an adult comes to mind.

It was Subrahmanyan Chandrasekhar's doctoral advisor, Ralph Fowler (1889–1944), who figured out that the white dwarf gravitational support mechanism could come from combining two basic principles in quantum mechanics: the Pauli exclusion principle put forth by Wolfgang Pauli (1900–1958) and the Heisenberg uncertainty principle put forth by Werner Heisenberg (1901–1976).

The *Pauli exclusion principle* prevents two fermions from having the same quantum state. Think of it like two people not being able to occupy the same place at the same time, no matter how hard they try to or how much fun they might have doing it.

The *Heisenberg uncertainty principle* limits the precision to which we can determine the position and momentum of something. In its simplest form, momentum is the mass of a particle times its speed. The very act of trying to measure the position of something as small as an electron would influence its momentum. It's like being in a completely dark room and trying to determine the position of the colored balls on a pool table by bouncing the white ball back and forth and listening. As soon as the white ball strikes a colored ball on the table, that colored ball moves to a different location and the direction of the sound becomes useless.

Electrons are fermions, so when they are compressed to the levels seen in a white dwarf, they exert a pressure simply by obeying Pauli's exclusion principle and Heisenberg's uncertainty principle. Electrons must have different quantum states, and the separation between the particles dictates their momenta, i.e., the speed at which electrons can zip about. When matter is compacted this closely, it is called *degenerate*, so a white dwarf is supported by *electron degeneracy pressure*. This pressure is limited by how fast the particles must be zipping around, and that can't faster than the speed of light. This is the limit found by Chandrasekhar and the one responsible for type Ia supernovae. Neutrons can also provide a *neutron degeneracy pressure* to support a neutron star. However, once this pressure is overcome, by adding even more mass, the only known object left in our astronomical arsenal is the black hole.

A large number of stars will end in the final equilibrium condition of a white dwarf. In these cases, quantum mechanics is balancing the force of gravity. White dwarfs are still giving off energy, though; otherwise, they wouldn't be white. This known energy loss rate means we can estimate both how old a white dwarf is and how long it will glow.

Given some estimated initial brightness, we find white dwarfs with ages of ten billion years that are still going strong[13]. These stellar remnants are the glowing embers of once-raging furnaces that are slowly being petrified as they move through lower and lower energy states. This push and pull between quantum mechanics and gravity keeps a white dwarf fair and balanced so that it glows on and on and on and on.

Part Three

The Future

Astronomy Saves the World

Chapter 8: Rock to Rock Exploration

Devastating impacts from comets and asteroids should never be something Earth has to experience again. Of all the known objects in the catalog of near-Earth asteroids, none of them has a score higher than a zero on the Torino scale. We cannot, however, say the same of comets, because these random events can be celestial surprises. We do have the ability to detect the worryingly large ones as they pass inside the orbits of the outer planets, and if we spot them early enough, we should have enough lead-time to redirect them. After all, there is only a twelve-year span between the Russians launching Sputnik, Earth's first artificial satellite, in 1957 and humans walking on the moon in 1969.

Much of the hard work has already been done. So, with the right motivation, a budget that isn't subject to the whims of politicians who are constantly trying to get re-elected, and enough time, deflecting the path of an asteroid or fast-moving comet should be achievable. There would be a lot of disappointed people if we failed to deflect an asteroid or comet, and it would be a little embarrassing if an alien species ever got around to visiting Earth and didn't find anything other than our residual junk still in orbit.

But being able to force rocks and dirty snowballs away from us does not mean we can relax if we want our species, or our descendant species, to survive into the deep future. Another popular disaster scenario is a rogue black hole passing close by.

As with many things, this threat is taken way out of proportion. Even though talking about a rogue black hole sounds cool, gravitationally, it presents almost the same problem as any other type of star passing close to, or entering, the solar system.

A close gravitational encounter with another star may increase the chance of impacts with rather large asteroids and comets as they get nudged out of their current orbits and into new ones. Although very unlikely, such an encounter may also nudge dwarf planets or moons onto an impact course with Earth. This would be something that we would have no hope of preventing, and is something that could have already happened. It is thought that a Mars-sized planet called Theia collided with the early Earth, and the moon is likely made up of the resulting debris[1]. The Theia scenario is consistent with findings from the Apollo missions, consistent with the observed orbital evolution of the moon, and something not unexpected in the violent beginning to our solar system.

It is worth remembering that space is still really big, and time is really long. So while close encounters with other stars have a very small, or zero, probability of happening in our lifetime, as time ticks by, the likelihood of an unavoidable impact goes up and up. The same is also true for the possibility of multiple impact events happening in quick succession. This could be just as disastrous. So, while we may be able to react quickly enough to deflect one of these objects, could we do two in the same year, or three?

What about a nearby star exploding and bathing Earth in massive amounts of potentially deadly radiation and particles? What about our own sun running out of hydrogen and expanding to the point where Earth is no longer habitable? There are no immediate worries about nearby stellar explosions, and the sun is not scheduled to kick us off our planet for at least a billion years. Both of these unavoidable events will happen at some point in the future, though. Moving Earth itself is a far-fetched, but not totally impossible, option[2]. Even if we did manage to move Earth, it would only be a temporary solution.

On top of the astrophysical problems, our planet is also facing global climate change and a currently increasing population. To prevent these from becoming even bigger issues than they already are, we have to find ways to fundamentally change our current agro-industrial fossil-fueled approach to our short-term sustainability and find new resources that can provide additional raw materials and more places to live.

These things considered, the argument for getting at least a few thousand people permanently, safely, and sustainably off this planet is compelling enough for us to do something significant about it right now.

There are detractors from the idea of establishing permanent off-world colonies. People surrender to a history that repeatedly shows our failures. However, I find the typical arguments against the expansion of our physical presence outside low-Earth orbit to be the shortsighted groans of a semi-disorganized, boring, self-destructive, and defeatist species.

Success in space drives success on Earth. Down here, we are almost completely inefficient. Out there, we are forced to be as efficient as possible. Down here, we squander resources. Out there, we must use every resource to its maximum potential and recycle everything. Of course there are things that we need to spend money on to fix here on Earth, but we can go a long way to fixing down here by expanding out there. Space exploration and improving social services are not either/or endeavors. They are complementary, and we can do both.

Despite the ramblings of the political establishment, some of us must embark on humanity's next greatest adventure, and we can do so by hopping from rock to rock. The objects that have typically heralded the potential annihilation of our species may now be our tickets to an auspicious future.

So how can a sustainable human presence outside low-Earth orbit be achieved? To approach this question, we must first understand the major issues we face. These center on our need to have an environment that won't kill us. We need the right temperatures, pressures, and atmospheric mixtures. We need the right amount of sunlight for the right amount of time, the same shielding provided by Earth's atmosphere and magnetic field, and a probably similar gravity environment. We need a sustainable source of food and water, a waste disposal or recycling system, and to make sure we can fix everything, including the human body, when things go wrong. These are big problems to solve, and top people are working on them right now. But before we can begin our journey, we need to know where to go.

The Ultimate Destination

Quite simply, our ultimate destination is any other star system. One glaring problem with this is that we don't know precisely what resources might be available once we get there. The resources don't necessarily have to be a habitable planet, but they have to be *something*. So, before we begin on any voyages that far, we really must know which star systems are worth aiming for. Astronomy is the only science that can tell us that, and it does it with the aid of telescopes.

The challenge we face when using telescopes to look for resources in other star systems is that these resources will be very faint. They will be reflecting their host star's light and could have a very low glow if they are somewhat warm. Therefore, we require telescopes that collect as much of that faint light as possible despite us being blinded by the exceedingly bright inferno that is the host star.

Those of us that travel from west to east during the morning commute and east to west during the evening commute are familiar with this problem. It is difficult to see anything around the glare of the sun. Unfortunately aligned traffic lights are a good example. The glare from the sun prevents a driver from being able to see the color of the light. This effect is typically worse during the end of March and the end of September, when the sun is rising and setting fairly close to due east and west.

No matter how hard we try to prevent it, the light that passes through a telescope is bounced around so that what should be a point of light ends up spread out as a disk of light with fainter rings of light around it. We call this the *point spread function* or the *Airy pattern*, after George Airy (1801–1892) mathematically derived its features. The point spread function also includes features from imperfections in the telescope that create spikes, blotches, and speckles on top of the disk-ring pattern. What makes matters worse is that these speckles can either hide, or look like, the very resources we are attempting to find, and they can change due to subtle temporal variations in the telescope itself.

In the era of big expensive telescopes, it is easy to forget that there is plenty of exciting and rewarding science that can still be done with small or modest-sized telescopes. This is because to simply detect the presence of something, we don't necessarily need to spatially resolve down to very small scales every single time. If, over the course of the next few years, there are suitable advancements in technology that can mitigate the problems of spotting faint things next to bright things, for a few billion dollars, we could simply build a modest-sized, properly equipped space telescope and take some very deep looks at the nearest systems while coping with the blinding light of the host star.

It still may be several decades before we construct a telescope capable of determining the precise nature of resources in the nearest star systems. In the meantime, we can still deduce useful facts about these systems using methods other than direct imaging. One of the most satisfying developments in astronomy over the last ten years has been our ability to detect planets around other stars.

Astronomy Saves the World

The *Kepler Space Telescope*, named after the chap that showed planetary orbits aren't perfect circles, is a 1.4-meter-diameter observatory[3]. It is equipped with a massive camera that stared at a single patch of sky for about four years. Its main mission was to find stars that are subtly and regularly winking at us. A planet blocking out a little of a star's light causes the winking, and there have been thousands of winks seen to date. This is basically the same method used to calculate the astronomical unit from the transit of Venus across the disk of our sun.

One of the major discoveries we made using the *Kepler space telescope* is that planetary systems are common around most stars. Therefore, even though we are currently struggling to look closely at the potential resources around our nearby stars, we expect most of them to have planets. An implication of the *Kepler* findings is that our solar system could be somewhat representative of every star system, i.e., we can expect asteroids and comets to be present no matter which star system we aim at. In addition, those comets and asteroids are presumably made from materials very similar to those found in our solar system. So while our solar system may be unique in terms of the numbers and configurations of our planets, it is not likely to be special in terms of the actual types of objects it contains. Therefore, it can—perhaps naïvely—be argued that any star system will contain the resources we need, and simply blindly heading for the nearest one is a perfectly reasonable plan.

As well as searching for resource-rich asteroids, comets, and moons and finding the most useful places on the surface of Mars, we must also study all the nearest star systems with our biggest and best telescopes as comprehensively as possible. Not only are astronomers responsible for understanding the history of our own origins; they are also our astrometric guides into the future.

Alpha Centauri, which is 1.3 parsecs or 270,000 astronomical units away, is the nearest star system and it contains two fairly sun-like stars and a red dwarf. The red dwarf, Proxima Centauri, is the closest of the three. There is some good evidence to support the presence of a rocky planet around Proxima, but regardless of the confidence in this data, it is unlikely that any system would be totally devoid of useful resource. We just need to keep looking[4]. If our own solar system is littered with watery rocks, why wouldn't other similar star systems be as well?

Getting There

If it turns out that the Alpha Centauri system is in fact the best place to head for, how would we go about getting there? Should we try to take another giant leap for humankind and get there as quickly as possible, or should we take our time and slowly and cautiously hop our way out there?

The giant leap approach would be to build a crewed spacecraft capable of huge speeds and simply aim it at Alpha Centauri. The obvious issue here is we don't yet have a vehicle that can safely make such a journey. The *New Horizons* mission hurtled past Pluto in the summer of 2015 after spending only nine or so years in transit. At *New Horizons*' speed, it would take tens of thousands of years to travel the distance to Alpha Centauri. There are, however, ambitious plans by the likes of Stephen Hawking, Russian entrepreneur Yuri Milner, and Facebook founder Mark Zuckerberg to accelerate micro-spacecraft to huge speeds using lasers[5]. This approach, assuming the necessary technology development takes place, may get something from Earth to zip through the Alpha Centauri system in about twenty years. Including the time it would take to develop and launch such a mission, it will be at least thirty years before we would see the tiny amount of data such a mission could collect.

The time it would take for us to get to the next star system in one shot presents an interesting thought experiment. Let's say we suddenly realize we have a way to make an engine that can get a crewed mission to Alpha Centauri in 150 years rather than twenty thousand years. In one hundred years, it will be a one-hundred-year-old piece of technology still fifty years away from completing its mission. The technology used will be the Ford Model T to what will become the Ford GT. During its long flight, that spacecraft will become obsolete. There is a significant chance that a faster spacecraft would be built in the meantime, be launched years after it, and still arrive at Alpha Centauri beforehand. In this case, what is the point of first trying to directly send a relatively low-tech spacecraft to another star system in the first place? Perhaps there is more to the slow and steady rock-to-rock approach than we first thought.

Hundreds of thousands of years ago, one of our ancestral species, probably *Homo erectus*, migrated to the point where the exploration of more land couldn't happen without them crossing water. When they reached this point, they did not set out for a new continent by clinching to a few fallen trees and randomly floating for a few thousand miles. Well, perhaps some of them did, but they were assuredly among the first recipients of the Darwin award. It's more likely that they saw a nice-looking, resource-rich island not too far from the shore and decided to swim to it.

With more experience and more ambition, they would have set their sights on another island even farther away. Sooner or later, the next island would be too far to swim to. Necessity is the mother of invention, so they must have built a boat to carry them across the distance. This leap forward in technology occurred, and the first genera of hominids were able to disperse across increasingly large stretches of water.

This island-hopping approach worked in the past, so why wouldn't it work now? We don't need to build a boat to cross the ocean—we just need to swim to an island, then the next, and the next. The technology will advance along with us, and sooner or later, we'll have the equivalent of an ocean-crossing liner. Perhaps, one day, we'll even have a space-*Concord*.

Earth is our island in space, and many smaller islands surround us. Several of us have managed to swim to the nearest large island, the moon. Some of us, tragically, didn't even make it into the water, and no doubt, there will be more casualties in the future[6]. And while our presence aboard the ISS has given us a lot of experience in extended human space flight, it is still us sitting in the surf and looking at an island our fathers already swam to.

Moon Base Alpha

Our lunar activities will be the dress rehearsals for Mars, dwarf planets, moons of the gas giant planets, and beyond. The six trips we did take to the moon, the last one being in 1972, were completed using equipment that was very low-tech as compared to today. It's vinyl versus online streaming. It's the steam engine versus Tesla Motors. Back then, the technology we have today was barely conceivable, but the basic physics of getting from point A to point B has not changed. Despite the technological leaps we have made in the last fifty years or so, we still need big rockets to get anywhere in space.

A good benchmark for the performance of the upcoming heavy-lift launch vehicles is that last crewed moon mission in 1972, Apollo 17[7]. In this case, the *Saturn V* vehicle successfully delivered fifty-five hundred kilograms of material to the surface of the moon. This included the equipment necessary for the wellbeing of the crew, their return to lunar orbit, and their link up with the command/service module that would return them to Earth. However, to establish a permanent presence on the moon, or any solar system body for that matter, we don't need to send any crew with the first mission. We can simply send the equipment needed to begin construction of a base. No, this is not science fiction. We can, if the right people or corporations were so motivated, start building a moon base almost immediately.

The construction of a moon base will be a job for the first lunar engineers, but astronomers are going to be the ones deciding where that base should be built. We need to know the availability of local resources and which location will provide as much protection as possible from radiation and potential impacts.

One of the favorite potential locations is the Shackleton crater at the south lunar pole[8]. The interior of some of the crater's walls are in eternal darkness, and other parts of the crater are in near-permanent sunlight. This presents a great opportunity for solar power. A buried or back-filled base would provide protection from impact debris and a shield against radiation. We have evidence that there may be significant amounts of ice hiding in permanently shaded regions of craters at the lunar poles[9]. NASA's *Lunar Resource Prospector* aims to conduct a detailed study of these craters, but, if approved, it won't launch until sometime in the early 2020s[9].

Assuming we find the right type of crater, can confirm the presence of water, and see the topography is favorable, the construction of such a base could happen in several phases over the course of, perhaps, a decade.

Phase 1 could be the first arrival of a robot team that would create the basic structures needed for a permanent base. Two good examples of this would be landing and launch pads free of dust and trenches for cabling.

Phase 2 could be the beginning of construction for the habitable part of the base. The technique of 3D printing could play a major role here. We've already seen large-scale 3D printings of cars and houses, as well as 3D printing taking place on the ISS. What we need to do now is figure out how to process lunar materials so that they can become 3D printer "ink."

Phase 3 could be the completion of the base by the first crew. While robotic technology is advancing at a delightful rate, presumably, there will always be some tasks that must be carried out by a human.

Phase 4 could herald the beginning of a permanent human presence on the moon. With ready access to water and an appropriate processing facility, we could expand the base so that a more comprehensive lunar facility could be established. The crew could have a more sustainable and comfortable environment and could work to make the base more capable in terms of manufacturing items from local resources.

An established moon base could serve as a refueling station for deep-space missions as well. If we can take it one step further and successfully process lunar materials into the materials needed to construct other spacecraft, or spacecraft components, then this will be a huge help. The delta-v to achieve lunar orbit from the surface of the moon is only 1.6 kilometers per second, compared to the ten or so kilometers per second needed to reach low-Earth orbit from the surface of Earth. Therefore, launching supplies from the surface of the moon is a lot less challenging than from Earth.

A moon base with manufacturing capabilities could become the stepping-stone needed to access the rest of our solar system. But, a moon base isn't going to solve all the problems associated with deep-space travel. If there were any emergencies, it wouldn't be a long wait before astronauts could get back to Earth or before help could come from Earth. It is also unclear whether the moon's gravity would be enough for an astronaut to overcome the medical problems associated with spending long periods of time off the surface of Earth.

Humans evolved with 1g of acceleration, and at the surface of Earth, g has an average value of 9.8 meters per second squared. It doesn't matter whether the acceleration is due to gravity or physical acceleration, and we could also say we have evolved in 1g of distorted space-time (thanks, Einstein). When we use the letter "g," typically, we are describing the force caused by acceleration relative to that created by the mass of Earth. We don't have much data on what happens to the human body when it's removed from the natural G-force it experiences on Earth for more than a year. So, it is unclear if longer missions are going to require some amount of artificial gravity.

Astronomy Saves the World

Valeri Polyakov spent 438 days aboard the Russian space station *Mir*. Apparently, he was "fine" when he returned to Earth. There have been a number of other Russians that have spent more than a year in space, and we now have data from when Scott Kelly spent almost a year weightless with his Earth-bound twin Mark Kelly as a baseline comparison. Commander Kelly apparently expanded by two inches in height during his mission.

While weightless on the ISS, astronauts spend around two hours a day maintaining their physical fitness. As part of their workout, they each run approximately three kilometers on a special Combined Operational Load-Bearing External Resistance Treadmill, or *COLBERT* (yes, this treadmill, seriously, is named after Stephen Colbert, the former host of Comedy Central's *The Colbert Report* who currently hosts *The Late Show with Stephen Colbert* on CBS). Without the physical exertion, the astronauts are simply floating around. Their hearts don't have to pump blood against the force of gravity. Their muscles have very little to do. Muscle mass and bone mass are both known to decline while weightless unless measures are taken to prevent the deterioration[11,12]. Understanding the negative effects of prolonged space travel on the body and how those effects can be mitigated is one of the ongoing studies aboard the ISS.

Much like the crew of the space station, a moon-base crew would need to maintain their physical fitness. This is simply so they can carry out the scope of work associated with getting a base up and running and keeping it that way. They would have far more options than those stuck inside the space station, though. The crew could, for example, purposefully wear mass belts around their waists, wrists, ankles, and heads to simulate the equivalent weight they have on Earth. This is only possible because there is some gravity on the moon, and it is a technique used by some people to increase their physical fitness in Earth's gravity.

Due to the relative ease (compared to Mars) of bringing people back from the moon, there are other ways to mitigate the potential medical issues. Each mission's duration could be kept to six or twelve months, and new crews could be swapped in and out just like the current routine on the ISS. In this way, they can simply be brought back to Earth and someone else can have a go.

Asteroid Mining

While establishing a permanent presence on the moon will provide massive insight into the medical effects of reduced gravity environments on humans (everyone may get taller but comparatively weaker), crewed missions to asteroids and comets are going to present their own set of challenges. These missions will have to battle the radiation environment found in interplanetary space, both from our sun and our galaxy, and potentially deal with high-speed collisions from debris, especially if chasing an active comet.

One way to overcome almost all of the challenges associated with a crewed asteroid mission would be to separate it into two phases, similar to those suggested for ARRM (NASA's proposed asteroid robotic redirect mission). Phase 1 could be a robotic capture and return mission. An uncrewed spacecraft would rendezvous with a suitable near-Earth asteroid, land on it, stop its rotation, and bring it back to an Earth orbit using the spacecraft engines. Care must be taken to avoid having the asteroid break into pieces, and some attention will need to be paid to the public relations issues of purposefully bringing an asteroid very close to Earth.

Once in Earth's orbit, an asteroid would be remarkably useful. This is where Phase 2 of such a mission would begin. Astronauts could use it as a training ground for future missions. If in low enough Earth orbit, the asteroid wouldn't be in the harsh radiation environment of interplanetary space any longer. It would be somewhat protected by Earth's magnetic field. This would reduce some of the potential radiation risks during the training. It would also, once and for all, demonstrate that we can protect ourselves from potential impacts involving these objects, which is ironic, since to demonstrate this, we would purposefully bring one *closer*!

To overcome the challenge of collecting the resources on the asteroid or comet, we could manually or robotically process the asteroid into more fuel, and other fundamental resources, making it so we would no longer have to launch such items from Earth. This would save huge amounts of money. Such a station would be ready to refuel any missions aiming to go onto other asteroids, comets, the moon, and Mars.

The moon base and asteroid mining missions are both different and complementary, and they could both become commercially profitable. But how much are they going to cost initially? It is relatively easy to assume that the establishment of a permanent moon base could cost somewhere in the region of 150 billion dollars (accounting for inflation). This is about the same as it cost to get the ISS to where it is today. An asteroid return mission is a bit trickier to estimate, as it hasn't been done before. *Rosetta*, which rendezvoused with, and sent a lander to, Comet 67P/Churyumov–Gerasimenko, cost 1.8 billion dollars, and *Hayabusa*, which rendezvoused and landed on the Asteroid Itokawa, cost about one hundred million dollars. The cost of *OSIRIS-REx*, the NASA mission that is planning to bring back a sample of Asteroid Bennu by 2023, will cost about one billion dollars. These numbers seem far too low to actually robotically return an asteroid to low-Earth orbit, so perhaps a conservative estimate for a relatively minor asteroid would range from five to ten billion dollars. A small price to pay for potentially saving Earth. Plus, the resources on the asteroid could be worth more than the mission costs in the first place. In fact, some companies are already banking on it [13,14].

Let those numbers sink in a little. They can be hard to really comprehend. Numbers that are five or six orders of magnitude outside our everyday experience are difficult to grasp, and so a number like ten billion may prove a challenge. It is four orders of magnitude greater than one million! However, we can simply contrast these numbers to some other "everyday" items that cost a lot of money. The relative benefits of these items might not be so obvious when compared to expanding the distant future of our species.

As of March 2016, the estimated replacement costs for the US Navy's Ohio-class nuclear submarines could exceed ninety-seven billion dollars. The development of the US Navy's latest generation of super-aircraft carriers has cost over thirty-six billion dollars, with the first of the line, the *USS Gerald Ford*, costing over ten billion dollars. *HMS Queen Elizabeth*, the super aircraft carrier, will cost about five billion dollars. The B2 Spirit stealth bomber program cost about two billion dollars per aircraft. These things start to add up, especially when we consider the cost of all the aircraft onboard the super-carriers. With the right motivation—fear—we can obviously build wickedly expensive projects. Given a bit of moderation in future military spending, on both sides of the Atlantic, doing things like bringing an asteroid into Earth's orbit and establishing a moon base should not be brushed aside by arguments of affordability.

The alternative approach to funding these species-enhancing projects is not to rely solely on government involvement. The mysterious free market could decide whether these things could actually be profitable. We now have private companies regularly supplying cargo, and soon humans, to the ISS, and if other resources are available to be exploited, then why wouldn't some companies, or individuals for that matter, consider taking on the challenge?

The world's largest companies make tens of billions of dollars in profits each year, and there are many individual multibillionaires among us. The top four most successful companies in 2015, according to Forbes, are all Chinese. Their combined *annual* profits exceed 138 billion dollars. Berkshire Hathaway (Warren Buffett's investment company) came in fifth, with JP Morgan Chase and Exxon Mobile in sixth and seventh place. Forbes also compiled the list of the most successful individuals, several having a net worth of over fifty billion dollars. Of course, it is not that simple, but if any of these companies, or individuals, could be convinced of the value of expanding our species permanently off Earth, then we'd be well on our way to getting there.

The Mars Problem

Let's say twenty or thirty years from now, we have a permanent presence on the moon and a nice-sized asteroid safely captured into low-Earth orbit. We'd be all set up on the moon producing lower-cost fuel. We'd also be producing fuel on the asteroid as well as knowing we can deflect asteroids to our benefit. We'll be awesome, but what should we aim to do next? As it happens, these things are already being considered, and two of them are particularly interesting.

Some have posed that we could hitch a ride on a long-period comet. Like a barnacle attaching itself to a whale for a free ride through the oceans, a spacecraft could catch up with a suitable comet, put itself in orbit or land, and get a ride to the outer solar system. It is worth noting that catching a comet costs about the same delta-v as putting something on an outer solar system trajectory anyway. However, we'd likely run out of resources way before we got very far, so the advantage of using a comet to travel means that we could potentially use the resources on the comet along the way. It'd be a massive fuel can.

The right type of comet could take a long time to come along, though, so in the meantime, we can focus on the second big aim, which is getting ourselves to Mars. It is the nearest planet on which humans might be able to survive. To get to Mars, we have to raise our space game by at least an order of magnitude. There is a reason humans haven't visited Mars yet—it is really, really, really difficult. Much more difficult than visiting the moon. There are a huge number of issues to overcome that will require hundreds and hundreds of teams, thousands and thousands of people, some new ideas, new technologies, and probably the cooperation of the whole world.

Successfully getting to the vicinity of Mars, from Earth or the moon, will depend on thousands of mission-critical issues. Most of these will derive from several key mission parameters. Perhaps the most notable is how long the proposed mission will take. This will depend on how long the mission directors and crew are willing to spend traveling, how long the crew can physically survive, and how much energy the mission has available. The fastest possible way to move between two points is to take the shortest distance, which would put the trip to Mars at about a month. This would require a massive amount of energy, though.

The most energy-efficient route between two points in the solar system is to send the vehicle on an elliptical transfer orbit around the sun. To transfer between Earth and Mars, the two planets must be appropriately aligned before departure, and we must also take the orbital motions of the planets during the travel time into account. The aim point of the vehicle must be for where Mars will be at the end of the trip. Both must arrive at the same point in space at the same time.

When a vehicle is following an elliptical transfer orbit, we can use Kepler's laws to determine its motion. Wolfgang Hohmann (1880–1945) proposed such an orbit in 1925, and now these types of paths are known as *Hohmann transfer orbits* [15]. For Earth to Mars, the transfer takes about eight or nine months, and the planets are aligned appropriately once every two years or so (every synodic period). If the astronauts want to return to Earth, they will need to spend just over a year on Mars before the right Mars–Earth orbital configuration is set for another nine-month-long return

Hohmann transfer orbit. In this type of mission, the crew would have to endure almost three years in a low-gravity (or zero-gravity) environment, bathed in the radiation from our own sun and the other harmful radiation producers in our galaxy, and be somewhat self sustainable. These are major problems, but not the only ones.

A flavor of the things that will be mission critical, to both the time in flight and the time on the surface, include appropriate fueling, appropriate living conditions, the processing of carbon dioxide, the production of oxygen, a sustainable power system, a heating and cooling system, appropriate management of water, appropriate management of food, radiation shielding, meteoroid shielding, artificial gravity and/or a comprehensive fitness regiment, a team of cross-trained people (perhaps six), a communications system, a navigation system, a maneuvering system, emergency fuel, emergency supplies of all other expendables, backup systems, the ability to maintain and repair all the systems onboard the spacecraft, the ability to maintain and repair the human body with limited resources, and the ability to manufacture a significant amount of the items needed for the mission to be a success. Should any one of these things fail, be miscalculated, or generally go wrong, it could result in a quick or drawn-out death for the crew.

Any interplanetary trip poses its own set of problems. There is no atmosphere, no magnetic field, no useful gravity, probably no fresh supplies, and a limited number of spare parts. The situation isn't that much better on the moon, where there is no useful atmosphere, no useful magnetic field, and only 0.17g. The moon does pass into Earth's magnetic tail for a few days each month, but the rest of the time, it is mostly exposed to the same solar wind that interplanetary sojourners will experience. This places a special emphasis on moon-based operations. There will be some technologies that can be tested and/or developed on the moon to help our interplanetary ambitions.

Like the moon, Mars doesn't have a protective magnetic field, but due to its distance from the sun, the solar wind will have less of an impact. Mars also has significantly less gravity at its surface, where we would have only 38 percent of our Earth weight. Unlike the moon, Mars probably has water available in a lot more places, and it does have a very thin carbon dioxide atmosphere. This means there is a wider variety of choices when we consider establishing a base on Mars, and a wider variety of choices when looking to the natural environment for possible support. But even with these differences, permanent operations at a moon base are going to be great practice for Mars.

As with most things in astronomy, the biggest problems arise from the massive distances involved. This is the same for a crewed mission to Mars. Due to its elliptical orbit, Mars can be between 1.4 and 1.7 astronomical units from the sun, which is much farther than the moon. Then there is the energy needed to get there. After the ten kilometers per second delta-v to get from the surface of Earth to low-Earth orbit, and the 4.3 kilometers per second delta-v to get into a low-Mars orbit, the delta-v to reach the Martian surface is another 4.1 kilometers per second. This is quite considerable and needs appropriate descent and ascent vehicles. Unless these are transported to Mars ahead of time, they will need to be built in situ. And then there's all the fuel to think of. This is where the moons of Mars may become important.

Two of the many interesting features of Mars are the moons Phobos and Deimos. Phobos is closest at about six thousand kilometers away from the surface. It is also the largest and has a diameter of about twenty-two kilometers. Deimos is at a much greater distance, about 23,500 kilometers, and it is smaller with a diameter of about twelve kilometers. Both are very similar to asteroids, which makes sense, since they both may have been at one time, and they could both be used as service stations. As well as the processing of any potential water into fuel, parts of descent vehicles, like heat shields, could be manufactured there. So, once the low-Earth orbit to Mars vicinity transfer is complete, Phobos and Deimos might be good staging areas before a visit to the Martian surface itself.

Just like getting off the Martian surface presents a problem, because we need a moderate-sized rocket, getting to the Martian surface from Mars' orbit is also a problem. There is a lot more gravity on Earth, as well as a moderate atmosphere. This means that if we want to land a spacecraft on the surface of Earth, we can use the atmosphere as a brake. The friction between the spacecraft and Earth's atmosphere slows entering vehicles down. The friction is also why we need heat shields on the spacecraft if we want it to survive. For the space shuttles, the heat shields were a network of black replaceable silica tiles. For Apollo-like capsules, ablative heat shields are vaporized and followed by a parachute system.

On the moon, heat shields are irrelevant, since there is no atmosphere. There also isn't much gravity. Landing on the moon is a comparatively tranquil affair that can be controlled by a relatively small hypergolic engine. A lunar decent module, as it did with Apollo, could even hover over the lunar surface for a handful of seconds if the original proposed landing site looks dodgy.

Mars, on the other hand, has a respectable amount of gravity but not much atmosphere. Some atmospheric braking is possible, but we need much larger heat shields and parachutes to get the necessary deceleration to land safely. The descent vehicle will also need to use a significant amount of thrust for the soft landings needed to avoid damaging the surface systems necessary for a Martian colony. Whatever fuel the landing vehicle has will likely be used during landing, and there may be none left for a later liftoff. So, if we want astronauts to return from a surface mission to Mars, we are almost left with no choice but to refuel the launch vehicles in situ. But first, we need to get there.

Astronomy Saves the World

When we pause and take stock of all the problems that must be overcome for a crewed mission to Mars, it quickly becomes apparent why it hasn't happened yet. Perhaps the biggest question is the one about how much stuff really needs to be taken. We know there are resources on Mars, so knowing how much to take isn't as straightforward as one might think. In situ resource utilization (ISRU) could include harnessing Martian water, minerals, metals, other useful chemicals, and the surface "dirt" itself to help make fuel, grow food, and build structures. This means that before a comprehensive mission design for a crewed trip to Mars can be proposed, we must figure out exactly how big of a role ISRU can take. There is an *ISRU quotient*. A quotient of one means the mission needs to take everything from Earth. A quotient of zero means the mission doesn't need anything from Earth. It is likely that the first mission to Mars will have a quotient greater than one because of all the setup equipment needed, but as operations on the planet continue, the quotient would evolve toward zero. At this point, the Martian colony would be totally self-sufficient, and our species could survive until the sun evolves too far no matter how bad things got on Earth.

Finding solutions to the Mars mission issues is clearly not going to be the work of astronomers. It must involve the full range of scientists, biologists, chemists, physicists, astrophysicists, and mathematicians. It needs a massive amount of aerospace, chemical, mechanical, and systems engineering. The mental health of the crew will be a critical component of a mission-focused team in routinely life-or-death situations, so psychologists need to be involved. The crew will spend extended periods in confined spaces, so they are going to need to like the place and have plenty of distractions. Providing for these needs could be the role of architects, interior designers, and artists. And of course, this is all going to cost a staggering amount of money from both the federal and private sectors. Therefore, the business and logistical management is going to be fundamentally important. The Mars problem brings together almost every skillset we have and uses them toward a common goal for the betterment of humanity.

The designing of a spacecraft is one of those fun daydreaming-type activities that many people have spent time thinking about. In science-fiction books and movies, we frequently find vehicles that include a huge rotating ring. The idea here is to use the rotation of the ring as an artificial gravity environment. This would mitigate many of the adverse physiological effects that a human would experience in weightlessness. The amount of radial acceleration (artificial gravity) that can be produced is equal to the rotational velocity squared divided by the radius of the ring. Earth gravity can therefore be produced on the inside of a hundred-meter-diameter circular ring rotating once every fourteen seconds.

One could argue either way about the feasibility of such a spacecraft. On the one hand, the ISS is about one hundred meters in diameter, albeit mostly solar panels, so building something that large should already be manageable. On the other hand, getting it to rotate that fast is going to require a lot of energy in the first place, and the astronauts onboard—if they decide to look out a window—may quickly become disorientated as their view is quickly spun around and around. The astronauts may also feel a little strange, as the acceleration that they'll feel at their heads, which are at a smaller radius of the overall ring, will be 5 percent lower than the acceleration felt by their feet; they will be stretched somewhat. It will be the same feeling that someone would get as they approached a stellar mass black hole.

Such a ship would need a series of separated but joined compartments making up the ring, and each compartment may need to be double skinned to reduce the possible damage from meteoroid impacts. If a meteoroid punctured one compartment, the damage could be isolated from the rest of the ship. Two compartments would need to be heavily shielded for when the radiation environment became too harsh due to solar or galactic activity. A heavily shielded compartment would have a lot more mass, and so at least two would be needed in order to maintain the balance of the rotating structure.

On a hundred-meter-diameter rotating ring, there should be plenty of livable volume. But more volume means more complexity, and more complexity means more things could go wrong. For example, such a large structure would have significant power requirements. Massive solar panels could be deployed in the space inside the ring, and they could also act as a crude solar sail. But the requirements will almost certainly need to be supplemented with nuclear power. Parts of the vehicle must also be detachable for emergency and landing purposes.

Regardless of what vehicle designs and surface systems are finally used for a mission to Mars, the opportunity to think about these things sparks the imagination of so many people, and that can be a great experience leading to great things. We have already seen impacts to life on Earth from the things needed to support a Mars mission. Indoor farming, water conservation, waste recycling, and general sustainability are all examples of things that have to happen on Mars but are a huge benefit here on Earth.

Beyond Mars

Let's move forward to the mid- to late-twenty-first century and assume we've cracked the Mars mission. We have a sustainable outpost on the red planet, an ISRU quotient approaching zero, and perhaps a support station on one or both of the tiny asteroid-like Martian moons, Phobos and Deimos. What's next? Heading for the next star, trying again to hitch a ride on a long-period comet, or more rock hopping?

In the grand scheme of things, Mars isn't really any closer to the next star system than Earth, and there are no significant advantages in trying to hitch a comet ride from Mars. Continuing our hops from rock to rock is the only practical way to make steady progress.

Astronomy Saves the World

The main asteroid belt, which lies between Mars and Jupiter, sounds a lot more impressive than it really is. There are billions of rocks, some icy, spread out over a ring with an inner edge of two astronomical units and an outer edge of 3.3 astronomical units. Some quick calculations show that even with a billion rocks, the inter-asteroid gaps are still millions of kilometers. This is very different from the asteroid fields written into both the *Star Trek* and *Star Wars* universes.

Even if all the masses in the main belt asteroids were added up, there would only just be enough stuff to make an object one quarter the mass of Pluto. The most massive asteroid is the 940-kilometer-diameter Ceres. This asteroid has been shown to contain water, but being only one-quarter the size of our moon, its surface gravity is only 3 percent that of Earth. Nevertheless, resource-rich Ceres makes for the most logical next step after Mars.

After Ceres, and perhaps a few other useful asteroids, it would be on to the moons of Jupiter. One would think it would be a quick trip onto the moons of Saturn, but it is frequently forgotten that Jupiter is only a little over halfway to the famous ringed planet. Humans setting out on a mission to Saturn from a moon of Jupiter would be in for the longest voyage yet. Uranus and Neptune would be next. Each time, the journey would be longer, and each time we reached a new place, we'd stop, set up shop, process the local resources, set up a permanent colony, and plan the next hop. Perhaps Pluto, or another Kuiper Belt object, would come next. By now, the hops would probably be crewed by people not born on Earth, and there is likely no way that people on this voyage would ever visit Earth.

We have now cataloged several thousand members of the Kuiper Belt. It is littered with icy rocks, some almost as big as Pluto, that will contain the resources we need to continue hopping farther and farther. It is thought that there may be hundreds of thousands of these mini-ice worlds stretching away from our solar system. Much like the main asteroid belt, these too will be separated by millions of kilometers. Indeed, now that the *New Horizon* mission is done with Pluto, it is heading for another Kuiper Belt object. It is scheduled to fly by 2014 MU-69 on New Year's Day 2019. Hopefully, this object will have a much more appealing name by then, but it will provide us vital data on the potential usefulness of these distant mini-ice worlds.

Once we get out to the Kuiper Belt, the travel time between stops will be less, but the density of resources may be lower than those found on the moons of the gas giants. The Kuiper Belt objects are small, faint, and sparsely distributed, so we cannot hope to find all of them using observations taken by current instruments. To truly know where all the resources are, humans will need to continue doing astronomy as we hop from rock to rock. Sizeable telescopes at the distance of Pluto are going to have much more success searching for new Kuiper Belt objects, and their counterparts in the Oort Cloud, than we likely ever will from Earth.

As discussed in Chapter 2, Jan Oort put forward the potential existence of this isotropic distribution of icy rocks, as well as many other groundbreaking contributions to astronomy, based on the observed orbits of long-period comets. These objects can have semi-major axes of over a hundred thousand astronomical units, a significant portion of the distance to the next star. This leads us to the supposition of a cometary reservoir at huge distances from the sun, distances much greater than the observed Kuiper Belt. While we haven't positively identified any of these Oort Cloud objects in place yet, they may be used as stepping-stones to the next star system. We've already seen the presence of Kuiper Belt-like regions around other stars, so it is possible that the Oort Clouds of neighboring star systems overlap.

We are now running into an increasing number of what ifs, but these aren't so farfetched as to render the plan to hop from rock to rock completely unfeasible. It is clear that water is relatively abundant, and that our solar system is not entirely unusual, so once we've gotten out to the Oort Cloud, all we'd need to do is carry out the reverse journey as we expand ourselves into the next star system. An Oort Cloud transfer would be followed by aligning with the orbital plane of the neighboring system, getting into the exo-Kuiper Belt, down into other rocky bodies in the outer edges of the star system, and potentially onto any interesting Earth-like planets that may be in the habitable zone. At this point, we'd truly be an interstellar species.

When our species first started to disperse from the African savannah, for whatever reasons—changing climate, dwindling resources, curiosity—they were heading out into a new frontier. The migrations went out in all directions, spanned many generations, and resulted in some genetic diversity as a result of the new environments in which people found themselves. New situations forced new ideas, and the good ideas helped propel the quality of life forward. There were no budgetary issues and no profiteering, at least for a while. Things, on the whole, got better and better and better, and what were once simple outposts with basic amenities are now thriving cities in defined countries that mainly have useful manufacturing capabilities.

We are now facing a scenario similar to the one faced by our ancestors thousands of generations ago. Today's dominant methods of energy production are doing more harm than good, so the climate of this planet is either on a knife's edge or already over it. Our planet's own resources are being squandered. Too many people unnecessarily suffer in poverty. Social unrest seems to be increasing as more polarizing ideas are shouted at the uninformed masses to produce wider cracks in our global adhesiveness. If current trends continue, we are heading for a bleak future. This is a very complex situation without a single simple solution. However, we have found part of the solution, and we've known it for thousands of years—the answer's in the sky.

Getting off the Ground

The achievement of sending humans to the moon and having them returned safely to Earth was met just seven years after the challenge had been officially issued. This moment in history briefly brought our species together to celebrate what we are able to accomplish, motivated an entire generation of the most powerful nation on Earth, and has inspired scientists and engineers into lifelong careers. Where is that motivation and inspiration today?

Our universe operates on timescales within which the existence of our species bears no relevance. The vastness of time is something common to us all and should throw the lengths of our lives into sharp contrast. This is fundamentally humbling, but it can also exhaust motivation and inspiration for the long-term good when, in fact, it should do the opposite. These timescales bring forth avoidable and unavoidable disasters, like rocks that should be resources rather than threats and a sun that will evolve to render Earth uninhabitable, but they also encourage very little forward thinking on a per-generation basis. "That will happen after I'm gone, so someone else can deal with it!" Without understanding the genuine motivations or the seriousness of the problems we face, the solutions may not be appreciated or supported. The necessary social and financial resources may be difficult to muster. So, if we are to get the next phase of human exploration off the ground, something must change in the minds of the masses. This is where astronomy must play a central role.

Despite the cognitive challenges, not everyone is ignoring the issues. Some progress is being made toward redirecting asteroids and getting people back beyond low-Earth orbit. But the progress is slow, and the solutions will certainly take longer than seven years. Indeed, the engineers and scientists working on these problems today will almost surely retire before they find the answers, passing their work on to new generations. Like the building of the Pyramids and the Great Wall of China, the efforts needed to secure a safe future for our civilization will span timescales greater than the current average life expectancy. Our next adventure will not be the giant leap that got humanity to the moon. It will be a series of small steps taken by a cohesive global society over the course of many generations.

Even in the presence of continually inspired billionaire philanthropists, these multi-generational problems will require the cooperation of multiple stable governments with continually strong leadership. If we are to be successful, the messages of this leadership must not focus on the time needed and the costs involved but on the value of the outcomes to society as a whole. The famous extract from President John F. Kennedy's Rice Stadium moon speech in 1962 can no doubt be recited by hundreds of millions, if not billions, of people: "We choose to go to the moon in this decade and do the other things, not because they are easy, but because they are hard." But it is the continuation of President Kennedy's dialog that has a deeper value. He went on to say, "Because that goal will serve to organize and measure the best of our energies and skills, because that challenge is one that we're willing to accept, one we are unwilling to postpone, and one we intend to win."[16]

We know that working on huge challenges like this has extraordinarily positive effects well outside the projects themselves, but the challenges we now face are far greater than landing a few men on the moon and having them return safely to Earth. "Procrastination is the thief of time"[17] so let's stop messing around, sort ourselves out, and get on with it. Missions to asteroids and comets, missions back to the moon, and missions to Mars will all rely on astronomy. We need to find where all the asteroids are and what threats and benefits they present. We need to find the most advantageous places to colonize on the moon and Mars. If we do not pursue these things, our time could be cut short, and the world could burn again, both literally and metaphorically. What would then become of our precious pale blue dot

Chapter 9: Emancipation via Education

Where would we be without astronomy? How different would life be if we lived on a completely clouded planet and had never seen the sun, the moon, and the stars? Day and night only differentiated by periods of hazy grayness, followed by almost complete darkness. Little help for predicting when warmer and colder times of year may be coming. No constellations visible from which to set a calendar, spark celestial imaginations, and encourage cosmic wonder.

Think of all that astronomy has taught us that is not obvious from a cursory glance: the age, size, and shape of Earth; the distance to the moon; the relative distances of the planets; the age of, and distance to, the sun, the other stars, and the nearest and farthest galaxies; the origins of the elements we are made of; the method of distributing those elements so our solar system could form; the age of our universe, the likely future, and our place in it; the threats to Earth and the opportunities they present; and the destinations we can look forward to. These are important things.

Our understanding of almost our entire origins, as uncovered by astronomy, compels us to cast aside ancient notions of supernatural causes. It has taught us that our species must mature if we are to overcome our own potential extinction. We are not acting out our lives following an intellectually subversive script in a privileged play on a cosmic stage. We have our own choices to make, our own future to nurture, and an unconditional responsibility for promoting the wellbeing of each other, both in the present and for the future generations to come.

There is still a lot of work to be done. There are still many of our species that, through their own circumstantial or purposeful ignorance of basic, universal facts, cling to superstition and myth. This is why it's so important for us to guard our education system against pseudosciences. If we succeed in protecting young minds from this nonsense, the number of people falling for these explanations will likely dwindle, and a more skeptical generation will come into focus.

In a world so dominated by the application of science, one has to wonder how so many people are still willing to capitulate their reason and regularly succumb to pseudo-academic flatulence. Every year, the same wild exclamations are made over e-mail or social media about a "super" moon (the moon is just a tiny bit closer to Earth than usual), about Mars becoming as big as the moon (impossible), about there being six continual days of darkness (impossible), about gravity being significantly reduced because of a planetary alignment (impossible), and that each of these things only happens once every 250 years or so. People also feverishly share memes about blood moons (which happens every lunar eclipse) and Mercury in retrograde (which happens every few months). Another claim boasts about there being five weekends in a particular month, and how that only occurs once every hundred years or so. Of course, a quick look at a calendar, or simple arithmetic, shows that five weekends in a month happens a few times each year.

Even a cursory knowledge of physics, astronomy, mathematics, and a dash of common sense should be enough for everyone to completely dismiss these types of claims. But something, scientific illiteracy or the knee-jerk reaction to believe almost anything that one reads or is told, ensures that these falsehoods are propagated time after time.

These types of cognitive errors may seem harmless, and many popular myths obviously give meaning to a great number of people. And that's fine, but only if these beliefs are kept personal, are not forced upon anyone else, and do not cause unnecessary suffering of anyone, especially children. Regardless of the weight people put on the meaning of myths, the myths themselves do not provide any useful explanations of anything that can survive even the most superficial of examination. So where is this examination? Why do people still succumb to these myths?

In many cases, current educational curricula place little importance on developing critical thinking skills and fail to encourage students to question things. Where are the common core exams on the scientific method? Where are the common core exams on basic philosophy? Where are the common core exams on the fundamental facts of our own existence? Slowly, we are getting there. The information age has allowed inquisitive minds to seek out this information for themselves and reach the level of understanding needed for the next great enlightenment. Even if we have reached a zeroth-order educational emancipation, there is still a lot further to go.

Intellectual Armor

The helplessness experienced by people who only have the opportunity to receive a basic education, if any education at all, means that many are forced to simply surrender their intellects to vitriolic rhetoric. In some parts of the world, an education system is almost completely absent, especially for women[1]. Many people are forced to obey the traditional rules being enforced by nations or communities frequently over-lorded by brutally repressive theocracies. In many cases, the members of such groups have no choice but to conform out of fear of capital punishment[2]. Human rights clearly suffer as a consequence. So, while there may be flaws in the Western education system, where overreaching theological bureaucrats compromise secular education now and then, at least some people will reach some point in their lives where they have the opportunity to freely think for themselves. In other parts of the world, this would be a huge struggle for many, many people at any point in their lives.

  Sadly, the gullibility of the masses is frequently used for manipulative purposes. The aim of the traditional peddler, or huckster, is to take advantage of someone and convince them they need something when they do not. In the process, that person makes some kind of profit. In many cases, the profits can be huge. Hucksters try to exploit ignorance and convince people that what they are saying is true, regardless of scientific facts. Their aims are to indoctrinate, take monetary or social advantage of someone, or both. These persons are typically in some form of leadership position within a group or are abnormally outspoken. A certain level of charisma is also typically present. They could be unwittingly taking advantage of people or doing it on purpose. They could have convinced people to part with a small amount of cash or donate a significant amount of money that people can't necessarily afford. They could have persuaded some people to buy their book or DVD or to visit their own themed attraction. They could have convinced some people to spread what they are saying to their children and other people, sometimes by force, or they could have convinced someone to detonate himself or herself in a crowded marketplace.

  I am often asked what I think of Deepak Chopra. Typically, this is because many people find this scholastic troll in some way profound. Chopra is a popular promoter of spirituality who used to be an endocrinologist. Oprah Winfrey likely caused his rise to celebrity after promoting his 1989 book *Quantum Healing: Exploring the Frontiers of Mind/Body Medicine*[3]. In his 2010 TEDMED presentation, Chopra claimed that people are unmoved by simple information, and their behaviors may only change for the better if they are subjected to an emotionally charged and engaging story[4]. From a psychological point of view, he might be right. However, just in the same manner that Chopra has no authority to speak on matters of quantum mechanics, I have no real authority to make any serious comments on psychology. Nevertheless, the sciences—and scientists—provide information without emotion, very rarely in a compelling way, and we struggle to change people's behavior. Hucksters, on the other hand, use the very tactic explained by Chopra to promote propaganda and bigotry and sometimes to incite violence. Many people would rather watch an entertaining bit of nonsense than spend the time needed to comb through dry data and come to a well-reasoned conclusion.

Televangelists are also renowned for making passionate appeals to emotion, and the billions of dollars they receive in donations is a clear testament to the success of this approach. The overwhelming majority of the tax-free and tax-deductible money donated to televangelists fails to find its way to charity and is shamelessly spent on personal extravagances [5,6]. This is not to say that the appeal-to-emotion approach is totally without merit. Many organizations will use a sad song and emotionally charged images of distressed children or young animals to elicit donations for genuinely worthwhile causes. But when the lifestyles of the most gluttonous televangelists are even superficially looked at, it should become clear that these people are hypocritical and revolting. They guilt gullible or desperate people into giving them self-crippling amounts of money and then use that money to buy multi-million-dollar mansions and private jets. Disgusting behavior. How many scientists behave in this way?

In these cases, the huckster is easily spotted and effortlessly dismissed by anyone that takes a moment to consider what is happening. But there are other, less obvious, hucksters in the form of the aforementioned Deepak Chopra. What sets his ilk apart are the attempts to appeal to intellectualism. Chopra has rightly received significant criticism for his public views, and there is no need to repeat the arguments against them here [7]. However, it is worth mentioning the manner in which he presents his opinions. In the same TEDMED presentation previously mentioned, Chopra goes on to make an appeal to emotion by claiming that it is possible, presumably through his magical mind/body medicine, to "reverse the biological markers of aging" by up to thirty years in only a few months. He made this claim while also ironically sporting a head of increasingly gray hair. Another example of a Chopraism comes from *The Enemies of Reason* documentary[8,9]. When asked to clarify his metaphorical use of the term "quantum theory," Chopra gave a completely nonsensical answer that tried to compare "creativity in consciousness" (having an idea) to a "quantum leap" (an electron gaining or losing energy and giving off light in the process). He went on to imply that because a thought is like a quantum leap, and a quantum leap is like biological healing (it is absolutely not), then you can heal yourself through thought! To me, it is gibberish. But to some, that might sound profound, and that is what Chopra bets on in order to make his money.

The concern with the approaches of the pseudo-intellectuals is that the points they raise in support of their own opinions are given more weight than they deserve, simply because of the academic tone of the delivery. When the substance of what is said is examined, we frequently find that nothing coherent is said at all. The confusability of their statements tempts their audience to simply bow to this fraudulent academic authority. One can sympathize with this reaction, because it is the one drummed into students from their first day of school: The teacher is always right.

Most people remember their favorite teacher. This is invariably because that teacher was genuinely good at teaching. A truly good teacher will encourage us to understand a complex subject with clarity. This demonstrates that almost everyone has an intrinsic passion for learning and that revulsion of learning can be promoted by the ineffective delivery of the material. Teachers are trusted to provide students with a *real* understanding of the world around them, which allows students to make informed decisions for the betterment of their own future and the future of others. A good teacher does not present arguments to purposefully confuse a student, and if a student is confused, the responsibility of that teacher has not yet been met.

By taking on an academic tone, pseudo-intellectual hucksters fake their way into a professorial position and lull the listener into the submissive student role. Do not be intimidated by a phony academic swagger. Claimants must explain themselves in a way that is understandable. If they cannot, the likelihood that what they are saying is literal nonsense is high. This is not to imply that quantum mechanics and general relativity are not a challenge to understand, but respectable scientists do not behave like the hucksters. Most do not even attempt to explain their research to non-professionals. But whether to a colleague or a member of the general public, scientists will generally point out the flaws in their own arguments, will rarely speak on things outside their area of expertise, and won't purposefully mislead people for their own financial gain.

In terms of honest intellectualism and the application of the self-correcting and academically rigorous scientific method, the rich history of astronomy provides an example to us all. Astronomers have struggled for knowledge and struggled to distribute that knowledge while also being faced with repeated totalitarianism. They were not hucksters hiding behind empty words and promises of increased wealth and wellbeing. They were intellectual heroes facing their data with open and critical minds, telling us how it is instead of how the majority of us wanted it to be. They did this despite what horrific punishment they faced at the hands of the ecclesiastical authorities. They uncovered the underlying majesty and beauty of the actual truth and moved us toward higher levels of understanding and technology. They fought, and still fight, to comprehend a counterintuitive reality as the cognitively lazy masses flocked to transcendent, egocentric, and extraordinarily self-contradictory explanations for their own existence. Astronomy has already saved the world, or at least some of it, from its own population.

The Fight for Knowledge

There will always be new discoveries to help us better understand the nature of our universe and compel more and more people toward science, engineering, and related careers. For example, we don't know the precise nature of dark matter or dark energy, we don't yet have a verifiable description of how gravity works on quantum scales, and we don't yet know precisely what is needed to spark life from organic compounds.

We are working hard to understand each of these things, and we have some ongoing hypotheses to keep the investigations moving forward. When we can confirm our understanding on the nature of the remaining unknowns, a Nobel Prize will probably be awarded, and scientists will congratulate each other on a job well done. No doubt, the media will have a frenzy for a few days before moving on to something else. It is also likely that the discovery will raise more questions than it answers and perhaps present some interesting implications. However, the first image of an Earth-like planet around another star, and the first data suggestive of life outside of Earth's biosphere, will be a whole different story.

The sciences, particularly astronomy, have been responsible for undermining dogma for millennia. Indeed, the scientific method of testing, retesting, and updating or discarding flawed theories is completely contrary to dogmatism in the first place. The flat-Earth, geocentric worldview is palpably false, as is that of a young Earth in a static universe. We occupy no special cosmic position and have no special cosmic privileges. The fact that the human race was not magically created out of dust or a clot of blood several thousand years ago is demonstrable.

Yet, there is one fundamental astronomical question regarding our place in our universe that has yet to be answered: Are we alone? Naturally, many people have many different opinions on this, ranging from the categorically crazy to the blissfully logical. The truth right now is that we simply don't know. If the answer turns out to be yes, then this planet, and we as a species, will be unique—some may say special. This would of course mean absolutely nothing more than we are incredibly fortunate. If the answer is no, then our understanding of our place in our universe would take the biggest leap toward insignificance that we have ever made. We would go through the latest and greatest cosmic demotion.

If, or when, astronomers demonstrate life elsewhere in our universe, it will be an absolute scientific triumph. It will mean that as humanity stretches itself away from Earth and hops from rock to rock, it will only be a matter of time before future explorers truly find a comfortable second home for our species. A new home with an appreciable atmosphere. A new home with liquid water available freely on the surface. A new home with easy-to-find resources. Hopefully any life we find won't be bad for our health.

Interestingly, we will never know with absolute certainty that life does not exist, or has not existed, somewhere else in our universe. There are simply too many planets that we will never be able to study in order to rule out the possibility. We cannot prove that there is no life anywhere else. We cannot prove a negative. However, we do know what to look for, we don't hide from the facts, and we've only really just begun searching in earnest for a second Earth.

The *Drake equation*, named after its proposer Frank Drake, is the most well-known approach to discussing the likelihood of life elsewhere. Its aim is to estimate the number of civilizations in our own galaxy with which we could communicate. There are two things that suggest the Drake equation was not originally intended to be taken seriously, though. The variables in the equation, essentially a string of probabilities, enable the estimated number of civilizations we could potentially communicate with to be anywhere from zero to several billion. Also, the quantity that it aims to calculate is not precisely directed at the question of whether there is *any* life at all elsewhere in our universe. A distinction must be made between the detection of a technologically advanced civilization and the simple forms of life that are likely to be a more common occurrence. Consequently, the Drake equation is only really relevant when searching for radio transmissions from extraterrestrial intelligences.

Sara Seager of MIT has purposefully updated the Drake equation. The *Seager equation* asks the more directed question: What is the likely number of planets that host any form of life that we will be able to detect? The answer is probably more than one.

We have very recently made significant steps forward in our understanding of the statistics that govern the probability of there being life elsewhere in our universe. The latest estimates are that at least one in five sun-like stars has a system containing an Earth-like planet capable of hosting liquid water[10]. Indeed, we are finding that liquid water is also possible under the icy surfaces of moons in our own solar system. This puts the number of potentially habitable places in our galaxy alone in the billions, and there are hundreds of billions of other galaxies. Therefore, the scientific consensus on whether there is any extraterrestrial life leans strongly toward probably.

We are now most definitely in the technological position to build an instrument capable of detecting life on another planet. We basically need a large space telescope equipped with instruments that can observe a planet's atmosphere and detect an overabundance of the gases produced by life. We can also estimate the approximate cost of such an observatory to be probably less than five billion dollars. There are hundreds of scientists, myself included, eagerly standing ready to carry out this project. All that is needed are the funds and ten to twenty years of effort.

Unlike the fundamental discoveries of the past, which were mostly made by individuals with little need for financial input outside their own means, the cost of the required observatory we need to spot an Earth twin leads to a potential roadblock. As has been experienced with the construction of the James Webb Space Telescope (JWST), the required budgets for space observatories can quickly get to the point where governments get directly involved. In the case of the United States, some government committee would probably have budgetary oversight and could cancel the observatory on a whim (as was the fate of JWST at one point[11]). Such a committee would almost certainly contain no scientists.

There is plenty of direct evidence to show how conservatism in the United States Congress repeatedly impedes the progress of science[12]. It does not take much to suppose the funding for an instrument that would attempt to answer such a fundamental question may be purposefully squashed in order to avoid the answer. Of course, there are other governments that could choose to fund such a project and many individuals whose net worth could mean they could take on such a project independently. Nevertheless, the scientists looking to construct such an observatory not only have to be well-respected experts in their fields but also have to be successful leaders of large projects involving a huge number of scientists and engineers, all on top of being masterful politicians. This is quite a broad set of skills that perhaps only a few people have.

Impact of Knowledge

If the signatures of life on another world are detected, what might be the response from the general public? Previous leaps forward in understanding our place in our universe, and the reactions thereof, serve as some kind of example that we could apply here. In the past, previous discoveries may have been slow to disseminate around the world and taught to only a privileged few. Everyday citizens did not have ready access to books, the Internet, or even news broadcasts. This led to only relatively incremental impacts. The discovery of life on other worlds could have an immediate influence and reaction much greater than we have seen in the past. Such a discovery will undoubtedly cause large numbers of people to seriously reevaluate their faith-based convictions and perhaps leave them behind. There will also be a spectrum of predictable responses to the contrary. Some will deny the evidence and fight against the facts, rather like young-Earth creationists. Some will distort the facts and twist them to fit new interpretations of old scriptures, rather much like the apologists. Some will try to show that the facts are not true and perhaps invent some ridiculous alternative explanation, like the conspiracy theorists. And some will simply ignore the facts and carry on regardless.

Considering that our current knowledge on the distribution of life-bearing planets is based purely on a sample of one (Earth), we can be confident that some massive, and possibly unexpected, steps forward in our understanding about the probability of life on other planets will take place in the next few decades. This is almost guaranteed by our current rate of increased scientific understanding and technological improvements and something that a lot of us are greatly looking forward to. We are going to learn a lot.

But we don't need to detect life on another planet in order to provoke the sense of wonder in our universe that drives people—especially young people—toward science, engineering, and mathematics. Sometimes, all we need is a groundbreaking, simple, yet powerful, picture.

Part of the beauty of astronomical images is that they are intrinsically profound. They do require explanations, or simplifying analogies, so that their relevance can be communicated to the public. Therefore, giving these pictures appropriate inspirational names and narratives becomes quite important. This argument is admittedly an appeal to human emotion, but it is an appeal that transcends borders and demographic backgrounds. It does not purposefully exclude anyone or encourage division. It encourages learning for all, and that can truly make a world of difference.

The *Hubble Space Telescope*, one of the sources of the awe-inspiring pictures, began its post-correction observations in late December 1993. Since then, it has produced a huge number of breathtaking astronomical images. The highlights for me have been the star-forming region in the Eagle Nebula, now known as *The Pillars of Creation*, the images of Jupiter showing the post-impact scarring of Comet Shoemaker-Levy 9, and the series of *Deep Field* images, which show how seemingly empty patches of sky contain hundreds of thousands of galaxies stretching back to a time when our universe was only a toddler. There is no doubt in my mind that these images played an important role in my decision to pursue astrophysics as a career.

In late December 1968, Apollo 8 emerged from behind the moon, and, for the first time in human history, a crew of three astronauts found the partially shadowed face of Earth staring back at them. The sun setting over western Africa, northern Europe in the grips of winter, Antarctica bathed in summer. Lunar module pilot Bill Anders, mission commander Frank Borman, and command module pilot Jim Lovell snapped the color picture that has been repeatedly called one of the most famous photographs of all time—*Earthrise*. This picture has been credited with providing the human race with a much-needed sense of global togetherness and humility during a time of significant social and political unrest[13]. Some people also claim that it may have started the environmental movement[14].

In 1972, the crew of Apollo 17, the last crewed mission to the moon, captured another iconic image of Earth. This time, there was only a hint of the day–night terminator, and the full sunlit side of Earth could be seen. This image, known as *Blue Marble*, is also credited with being highly influential, once again showing a small borderless oasis hanging in the vast blackness of space. This image provokes the senses of wonder and helplessness, and it has been replicated several times by various satellite missions over the past fifteen or so years. While these images are not new to us, these revisits at least serve as important reminders of the point they make. A widely distributed *Blue Marble* image came in 2015 from the Deep Space Climate Observatory (DSCOVR) satellite at a range of about one and one-half million kilometers[15]. Later in 2015, the same satellite observed another awe-inspiring sequence of images as the moon transited across the disk of Earth. People love the *Blue Marble*.

After *Earthrise* and the first *Blue Marble*, we had to wait another eighteen years before the next jaw-dropping image of Earth was presented to the human race. Not only is the image itself astonishing; this time, it came with the luscious commentary of Carl Sagan. *Voyager 1* took the *Pale Blue Dot* image in 1990 at a range of about six billion kilometers, i.e., about the same distance that Pluto is from the sun (forty astronomical units). This image was not part of the original *Voyager 1* mission image sequence, because it doesn't have any direct scientific value and could have resulted in damage to *Voyager 1*'s imaging devices. Nevertheless, it was Sagan himself who led the charge to convince NASA that there would be a broader value to such an image. While perhaps not as interesting as *Earthrise* or *Blue Marble* on first inspection, it doesn't take long to realize that the lack of detail is precisely why this image is so powerful. Sagan's commentary for this image is contained within his well-known book *Pale Blue Dot: A Vision of the Human Future in Space* published in 1994, two years before his death[16].

How many people have been inspired because of these images of Earth, and how many people may be inspired by a long-overdue similar new image? What will be the power of the first image of an Earth-like planet around another star? Such an image may be a significant step in convincing our entire species that either life does exist outside of Earth's biosphere or the possibility of life elsewhere should be taken very seriously.

In reality, the first data to suggest we may have detected life elsewhere in our universe will not come from a picture. It's most likely that it will come from the technique of *transit spectroscopy*, which is something the upcoming JWST will be capable of carrying out. The technique gathers the spectrum of a star during a planetary transit. A tiny fraction of that starlight will pass through the atmosphere of the transiting planet. The planet's atmosphere will absorb some of that starlight, leaving a trace of what it contains, and it's in those traces that we will be able to determine the molecules present in the atmosphere of the planet and whether they are indicative of life.

While transit spectroscopy may be the first method to seriously demonstrate extraterrestrial life, it will be with a very noisy spectrum that is difficult for the average person to interpret. What we will really need is a picture of that planet. We already have several dozen direct images of planets around other stars, but these are all more similar to Jupiter and nothing like Earth.

Is it conceivable that such an image may simply be consigned to the ever-increasing catalog of astronomical images rather than being tagged as the next *Earthrise, Blue Marble,* or *Pale Blue Dot*? Without actually having that image, it is hard to say. Even if we did have it in hand, it wouldn't be a direct image of us. We'd be looking at a metaphorical reflection of ourselves and the possibilities our universe presents due to the plurality of worlds. Would that make it *more* profound? Perhaps the power of the image will come with the slickness of the media delivery and silkiness of the commentary. Whatever the reaction will be, that first image of an Earth-like world is something that a lot of astronomers are working tirelessly to obtain. If the public doesn't yet know what a big deal it will be, the scientific community is certainly expecting it to have a significant impact.

Astronomy Saves the World

There is substantially more to our future than the development of technology for the re-direction and exploitation of debris in our solar system and for finding out which star system to head for. Astronomy is an educational gateway for everyone to understand basic mathematics and physics. It tells the compelling and demonstrably true story of our existence. It holds no preconceptions or orders about skin color, gender, sexual orientation, religion, or political opinions. It shows we are only one species on one planet, and that we have the opportunity to transcend Earth as we tiptoe into the relative unknowns of our home galaxy. This type of knowledge, if provided to everyone without exception, will certainly improve the quality of life across the planet. We could become truly united as we push the boundaries of what is possible for the first time in our short history.

Education is the key to emancipation from tyranny. Education is a basic human right. Everyone has the right to make informed decisions after being able to objectively review the evidence, especially on matters that involve their own bodies and their own wellbeing. Access to a free and secular education should be the primary mandate of any government that takes human rights seriously, and the law should protect the rights of anyone to get that education. In 1816, Thomas Jefferson wrote, "Enlighten the people generally, and tyranny and oppressions of body and mind will vanish like evil spirits at the dawn of day." [17] It will take a bit of time, money, and some strong politicking, but the time has come. It's time to enlighten people all over the world whenever possible.

We cannot successfully instill the type of STEM education we desperately need without the appropriate motivation. Some people naturally find it difficult to learn, especially when they cannot relate to the material. It is here that the majesty of astronomy takes hold once again. Most people love astronomy and want to know more. It helps provide a valuable education through exploration. When people learn what astronomy has to offer, they become more enlightened. They get to understand more about their origins and the significance of their home planet. They get to see how the odds are stacked against us, how the odds can be turned in our favor, and how we must battle these odds as one undivided species looking out for each other, forever, in peace.

The natural artistry that results from images of astronomical objects encourages us to learn more about science. We are the *Hubble Space Telescope* generations, driven to science by the corrected vision of the little telescope that could. Astronomy takes us from the absolute smallest scales to the yawning voids between clusters of galaxies. Our common misconceptions are fixed, and our appreciation for our own limitations installed. Astronomy inspires a love of learning, and the more we learn, the more we come to terms with logic, reason, rationality, and the power they have to change the world for the better.

Astronomy has a rich history, promotes science, provokes critical thinking, and is a subject that should be purposefully introduced into the secondary education system. Today, it is mainly left to a motivated teacher to make special efforts to include this material, or for students to form clubs and study it independently. We should change this. Astronomy should be part of a core curriculum.

If astronomy were taught to everyone, we could be a society of informed thinkers that would realize the value of knowledge and celebrate it rather than mock it. We could be a society of informed thinkers that would understand the value of reason and the scientific method. We could be a society of informed thinkers that would reject illogical, unreasonable, and entirely speculative worldviews promulgated by zealots. We could be a society of informed thinkers comprehensively emancipated from intellectual dishonesty because of a thorough, free, and secular education. We would be able to articulate our evidence-based knowledge against unreasonable arguments and defend ourselves against unsubstantiated proclamations. We can define our own reasons for being here, we can define our own moral standards, and we can define our own set of simple questions for establishing whether someone shares these values: Are human rights universal? Are evidence and reason more important than faith? Should people be prepared to change their minds? Yes, yes, yes.

Should we progress to a society where everyone values these things, it will be far from a utopia. There will always be challenges to overcome and people that need help. At the very least, accidents will happen, and decisive actions based on sound reasoning will need to be made. There are no comic book superheroes or otherwise supernatural forces to intervene. We've got us, and that's it, and that's okay. We obviously don't know everything, and that's okay, too. We at least realize the shortcomings of our knowledge and how to find out more. At least we are beginning to appreciate the value of a quality education, even if we don't yet know the best way to administer one.

There are clearly a great number of very talented students coming out of the secondary education systems of the world, who are going on to have very successful higher-education experiences and careers. Yet, the average understanding of basic mathematics seems to be slipping [18]. There are some simple principles that are either not covered or not absorbed by high-school students. The impact on higher education is that a growing number of freshmen must be mathematically remediated, because they struggle with basic algebra and trigonometry.

In my *opinion*, these slipping standards are not the result of bad high-school teachers (and I'm not just saying that because I am married to a teacher) but a symptom of a failing or at least grossly misdirected education system—most notably in the United States. I can only speculate as to the reasons for this, but I am impressed with the industrial education system explanation popularized by Sir Ken Robinson[19]. In this, our current system, we focus on the mass production of students that must meet a set of minimum quality-control tests, especially in STEM areas. Such a system does not recognize or encourage the development of a student's natural strengths and can force otherwise capable and talented human beings into careers they do not enjoy.

In such an industrialized model, it is easy to imagine the standard curriculum in the STEM subjects being watered down into a one-size-fits-all situation. In fact, in the No Child Left Behind Act's model, one need not imagine this at all[20]. It appears to be happening. It is no wonder that so many people despise mathematics and physics. It's also a tragedy that they may fight against learning some rather straightforward, yet powerful, ideas, techniques, and facts that could genuinely help them lead less stressful, and potentially more productive and fulfilling, lives. Imagine how different society would be if everyone were provided with a zero-cost secular education in subjects like astronomy for which they have a natural interest.

Imagine the kind of world we would live in if *everyone* understood that Earth is obviously not flat, is ancient, and is not the center of everything. Imagine the kind of world we would live in if *everyone* understood the origin of the elements that make life possible, and that our universe had a demonstrable and remarkable beginning. Imagine the kind of world we would live in if *everyone* carefully considered evidence, appropriately weighed facts, and took compassionate actions to improve our world as a whole. Imagine the kind of world we would live in if *everyone* knew the scientific method and how it forces us to correct our understandings of things when we are given new evidence. Imagine the kind of world we would live in if *everyone* had the skill to spot pseudoscience and the illiterate re-babblings of deluded money-grabbing despots trying to hijack moral compasses. Imagine a world where all of our generations think hard and work hard so that the next generations can build on our successes, learn from our failures, and take us even further into a prosperous future. What kind of world would that be?

Acknowledgements

Writing can be a lonely experience, but the writing of a book is not done alone. In this case, there has been a large team of people who have provided support and must be recognized.

The first are my wife and two children for their patience and understanding with the late nights, early mornings, solitary weekends, and frequent exhaustion-induced grumpiness. Now that I understand the pitfalls and pressures, if I ever do this again, I will manage the time needed very differently.

The second are two outstanding editors, Ashley Williams and Sabrina Leroe, who suffered through my rookie errors, gave me the benefit of the doubt, and provided candid feedback. Many people saw the early drafts of this manuscript, and the improvements from then to now have made a huge difference in the flow and clarity of the information presented. Not only am I deeply grateful for their efforts, but the reader should be as well.

Finally, I'd like to acknowledge the support of the crowdfunding community, especially those contributors that identify as secular. A significant portion of the funds needed to publish this manuscript came from the support of a Kickstarter campaign. That campaign received a much-needed boost after a podcast with the awesome Seth Andrews, and the many backers that provided some excellent feedback on the early draft chapters. To all backers, I'd like to extend my thanks to you, especially

Astronomy Saves the World

Angelia D. Anderson, Ian Batcheldor, David A. Bishop, Mr. & Mrs. B. Bomser, Mike Bowbyes, Bill Dreisbach, David Eisma, Peg Gantz, Chris and Tina Hoffman, Ron Lavigne, Isaac Silver, Susan C. Steinkraus, Jan Welke, and Jocelyn Marie Williamson. Special thanks go to the following backers for their incredibly generous support: Phyllis Albritton, Anthony Apfelbeck, Alan Batcheldor, Ruby Batcheldor, Sarah Batcheldor, Lori Hamilton, Dan Hardy, Nicolas Pain, Arnie and Sue Taub, Lindsay Wood, and all the people of MESATech. For all of the motivation you gave me to see this through to the end, I will be eternally grateful.

Glossary

absorption lines: Dark bands in spectra (rainbows) that show the presence of elements and molecules.

accuracy: How close an answer is to the truth.

ad hominem: A criticism of the arguer rather than the argument.

Airy pattern: The pattern around a point of light due to passing through a telescope. The pattern has a central Airy disk and is surrounded by increasingly faint Airy rings.

Alnilam: The middle star in Orion's Belt.

Alnitak: The east-most star in Orion's Belt.

analogy: A simpler comparable example used to explain a complex example more clearly.

Andromeda: The nearest major galaxy to our galaxy.

anti-intellectual: One who has a distain for genuine intellectual pursuits, particularly science.

apophenia: Seeing a pattern in random data.

apparent angle/diameter/size: How big something looks. This changes with distance.

arc length: A length expressed as an angle ($\theta = s/r$, $s=r\theta$).

arcminutes: There are sixty arcminutes in a degree.

arcseconds: There are sixty arcseconds in an arcminute.

asteroid: A space rock bigger than one meter across that hasn't hit a planet.

astronomer: A scientist that observes our universe.

astronomical unit: The average distance between Earth and the sun.

astrophysicist: A scientist that uses mathematics, physics, and chemistry to understand the nature of our universe and everything in it.

Big Bang: The event that resulted from the triggering of our universe.

binary stars: Two stars gravitationally bound to one another.

black hole: Any sufficiently small mass whose escape velocity is the speed of light.

blueshift: The change in wavelength of spectra from objects moving toward us.

carbon dating: Measuring the age of something using the radioactive decay of carbon-14. The half-life is around six thousand years, and this aging method is reliable up to about fifty thousand years.

Cepheid variables: Stars that regularly vary their brightness. The period of the variability correlates with their absolute luminosity, meaning they can be used to determine distance.

Chandrasekhar limit: The maximum mass for a white dwarf star. If the white dwarf exceeds this mass, it will explode as a type 1a supernova.

chondrite: A stony meteorite that formed at the beginning of our solar system.

CNO cycle: The nuclear fusion cycle of carbon, nitrogen, and oxygen that takes place in stars slightly more massive than the sun.

comet: An icy body that used to be in the Kuiper Belt or Oort Cloud. As they get close to the sun, the ices melt and make the comets' spectacular tails.

constellation: An area of sky that contains a well-known pattern of stars. Officially, there are eighty-eight constellations.

cosmic microwave background: The detectable signal from the Big Bang.

cosine: A trigonometric function that relates angles to lengths in a right-angled triangle.

cosmic rays: High-energy particles accelerated by exploding stars and black holes.

cosmology: The study of the origin of our universe and its structure on the largest scales.

Cretaceous: A geological period that ended at the same time as a giant impact from an asteroid.

dark energy: A poorly understood energy responsible for the acceleration in the expansion rate of our universe.

dark matter: A poorly understood substance that carries the gravitational force and keeps galaxies and clusters of galaxies together.

delta-v: The change in velocity needed to move between two points in space.

dimensional analysis: The method of checking the units used to see if an answer makes sense.

Doppler effect: The change in frequency of an emitter as it moves toward or away from an observer.

electron: A negatively charged fundamental particle about eighteen hundred times less massive than a proton.

electron degeneracy pressure: The force supporting a white dwarf star from gravitational collapse.

elliptical galaxy: A galaxy without spiral arms that has a smooth appearance.

equilibrium: When two or more forces are balanced.

event horizon: A region of space around a black hole where the escape velocity is equal to the speed of light.

Fermi method: The method that quickly determines a rough solution to a complex problem using limited data.

fission: The splitting of one atomic nucleus into two atomic nuclei. A large amount of energy is produced in the process.

flux: The amount of energy received by some area every second.

fusion: The joining of two atomic nuclei into one atomic nucleus. A large amount of energy is produced in the process.

galaxy: A collection of billions of stars held together mostly by dark matter.

gamma rays: High-energy rays of light harsher than X-rays.

globular clusters: A collection of one hundred thousand or a million stars in tight groups that orbit galaxies.

Goldilocks principle: When conditions are "just right" for an event or result to occur.

Goldilocks zone: The region around a star where conditions are just right for liquid water to be potentially on the surface of a planet.

GPS: Global Positioning System. It enables you to determine your position on Earth from satellite data.

granules: Blotches on the sun due to convection.

gravitational lensing: The distortion of light, due to general relativity, as it travels past a mass. Most noticeable around clusters of galaxies.

gravitational waves: Distortions of spacetime as a result of the motion of a mass through it.

gravity assist: Using the orbital motion of a planet to accelerate a spacecraft.

half-life: The time taken for 50 percent of a sample of material to radioactively decay into a different isotope or element.

Heisenberg uncertainty principle: On very small scales, there is a limit to how well the position and velocity of a particle can be known, because the act of observing the system disturbs it.

Heliocentric/Heliocentrism: The fact that the sun is at the center of our solar system.

helium: An atom that contains two protons.

hydrogen: An atom that contains one proton.

Hydrostatic equilibrium: When a pressure balances the force of gravity.

hypergolic: Two chemicals that automatically ignite when combined. Used for rocket engines once a vehicle is in the vacuum of space.

inferior/inner planet: A planet closer to the sun than Earth.

inflationary hypothesis: The leading explanation as to why our universe looks the same in all directions despite being incredibly large. It is also used to explain some other less obvious observations.

inverse square laws: Relationships in which one parameter changes as the inverse square of the other. For example, something that is twice as far away will be four times less bright.

isotope: Defined by the varying numbers of neutrons in an atom. Helium has two protons but can have one neutron (Helium-3), two neutrons (Helium-4), three neutrons, etc.

Joule: The unit of energy, i.e., the force acting on an object over a distance of one meter.

Kepler's laws: (1) Planets have elliptical orbits. (2) Planets travel faster closer to a star and slower farther away; their orbits sweep out equal areas in equal times. (3) $P^2 = a^3$. The orbital period squared is equal to the semi-major axis of the elliptical orbit cubed.

kinetic energy: The energy of an object as a result of its motion.

Kuiper Belt: A region that extends out to at least fifty astronomical units from the sun. It contains a large number of icy bodies that could become short-period comets, and a number of dwarf planets like Pluto.

Large Magellanic Cloud: Minor galaxy companion of the Milky Way one-hundredth its size.

logarithmic function: A mathematical function that changes rapidly because it is determined by the power given to a base number like two or ten.

logical fallacies: A faulty attempt at reasoning made so by a lack of rationality.

luminosity: The energy radiated by an object per second.

lunar eclipse: When the moon passes into Earth's shadow.

main sequence: The position in temperature luminosity space occupied by a star that is fusing hydrogen in its core.

Magellanic Clouds: Two dwarf galaxies in orbit of the Milky Way. They can be easily seen in the southern hemisphere sky.

Malmquist bias: The bias caused by not being able to detect faint things that are far away but being able to detect bright things that are far away.

mass-to-light ratio: The ratio between the necessary gravitational mass to account for the dynamics of a galaxy and the light observed coming from the stars in a galaxy. The ratio is determined by the presence of dark matter.

mass spectrometer: A machine capable of detecting the relative amount of different elements. Can be used in radiometric dating.

megaparsec: One million parsecs.

Milky Way: The galaxy to which our sun belongs.

Mintaka: The west-most star in Orion's Belt.

momentum: The mass of an object times its velocity. It is easy to stop a small thing traveling slowly but difficult to stop a large thing traveling at any speed and a small thing traveling quickly. Things with more momentum are harder to stop.

neutrino: A neutrally charged fundamental particle much less massive than an electron.

neutron: A neutrally charged fundamental particle about eighteen hundred times more massive than an electron. The number of neutrons determines the isotope of an atom.

neutron degeneracy pressure: The force supporting a neutron star from gravitational collapse.

neutron star: A star supported by neutron degeneracy pressure. The fate of main-sequence stars greater than about eight solar masses once they run out of fusible atoms.

Oort Cloud: A region that is thought to extend out at least one hundred thousand astronomical units from the sun. Likely contains a large number of icy bodies that could become long-period comets.

open clusters: A cluster of stars formed inside a galaxy as a result of the gravitational collapse of a giant cloud of gas and dust.

order of magnitude: Using powers of ten to distinguish between numbers, i.e., one hundred is an order of magnitude greater than ten.

parallax: Two different observing positions produce a shift in the apparent position of a foreground object against a background. The size of the shift depends on the distance to the foreground object.

parsec: The distance at which a foreground object shows a shift of one arcsecond when the distance between observing positions is one astronomical unit.

partial solar eclipse: When the moon passes in front of the sun but does not totally block it.

Pauli exclusion principle: Prevents two fermions (a family of fundamental particles, including electrons) from having the same quantum state at the same time.

photon: A particle of light.

Pi: The ratio of a circle's circumference to its diameter.

Planck's constant: A fundamental constant that links the energy of a photon to its frequency.

planetary nebulae: The nebulous ejecta of a recently deceased main-sequence star of less than about eight solar masses. The ejected material is illuminated by the radiation coming from the remnant white dwarf.

polarization: The orientation of a wave oscillation.

positron: A positively charged fundamental particle about eighteen hundred times less massive than a proton. The anti-matter version of an electron.

post-theism: The worldview that theism has been made obsolete by science and reason.

precision: Being repeatedly close to an answer.

proton: A positively charged fundamental particle about eighteen hundred times more massive than an electron. The number of protons determines the type of atom.

Proxima Centauri: The nearest star to the sun. Not visible with the naked eye.

Ptolemaic/geocentric: The incorrect view that Earth is the center of our universe.

quantized: When a system can only exist in discrete states.

quantum mechanics: A description of the counterintuitive actions of fundamental particles like electrons and photons.

radian: The description of an angle as defined by the arc length divided by the radius of a circle. There are 2 times pi radians in a complete circle ($\theta = s/r$).

radiometric dating: Using the half-lives of isotopes to date samples.

redshift: The change in wavelength of spectra from objects moving away from us.

relaxation timescale: The time taken for gravitational interactions to change the velocity of a star by about the same amount as its original velocity.

refraction: The deflection of the path of light as it moves from one medium to another. The size of the deflection depends on the wavelength of the light.

retrograde: When the apparent motion of a planet moves from west to east rather than the typical east to west.

rocket equation: An equation that relates the delta-v achievable by a spacecraft to the amount of fuel needed to make that maneuver.

sagitta: The minimum distance between the base of an arc and the center of the arc.

scientific method: The self-critical process of determining what is true, and the probability of something being true, using reproducible and measureable evidence.

shadow cone: The shadow behind a sphere or disk aligned perpendicularly to the rays of light coming from a source of light that is not a point, i.e., the shadow of Earth due to the sun.

significant figures: The number of digits that are most relevant to the value of a result.

sine: A trigonometric function that relates angles to lengths in a right-angled triangle.

Sirius: The brightest star in the night sky.

small angle approximation: When the angle of a triangle is small compared to the length of two sides, the curvature of an equivalent arc is insignificant, and the triangle can be treated like an arc. Here, the angle in radians is equal to the arc length divided by the radius ($\theta = s/r$).

Small Magellanic Cloud: Minor companion galaxy of the Milky Way but 140 times smaller.

solid rocket boosters: Powerful rockets frequently used in the initial ascent stages of a launch vehicle.

spatial distribution: How stuff is spread out around a place.

spectra/spectrum: The result of light being dispersed into its wavelengths.

spectral energy distribution: How the light from something changes with wavelength.

spectrograph: An instrument used to make a spectrum.

spiral galaxy: A galaxy with spiral arms.

STEM: Science, Technology, Engineering, and Mathematics.

STEAM: Science, Technology, Engineering, Art, and Mathematics.

strong force: A fundamental force that binds together the nuclei of atoms.

superior/outer planets: Planets farther from the sun than Earth.

supernova: The explosive end of stars with more than about eight solar masses, and white dwarfs that exceed the Chandrasekhar limit.

synodic period: The time separating two similar relative orbital alignments between Earth and another planet. This time is important when considering traveling to another planet.

tangent: A trigonometric function that relates angles to lengths in a right-angled triangle. Can be found by dividing the sine function by the cosine function.

temporal variations: How something varies with time.

theory: A provable scientific principle. Not a conjecture.

Torino scale: The level of threat an asteroid has for life on Earth.

total solar eclipse: When the moon passes directly in front of the sun and blocks out its entire disk.

transit: When a small object passes in front of a larger object.

transit spectroscopy: When the spectrum of a transiting object is detected. This is a powerful technique for examining the atmospheres of planets around other stars.

unit circle: A circle that has a radius of one.

variable stars: A star that varies its brightness in a cyclic fashion.

vector: A quantity that has both a magnitude and direction.

virial theorem: Relates the gravitational potential energy to the kinetic energy of a system. Can be used to determine total gravitational mass by observing dynamics.

watt: The unit of power, i.e., energy (joule) per second.

wave function: A mathematical function that describes quantum states.

weak force: A fundamental force that causes radioactive decay.

white dwarf: A star whose gravity is supported by quantum mechanics. The fate of a main-sequence star up to about eight solar masses once it runs out of fusible atoms.

WIMPs: Weakly Interacting Massive Particles that may constitute dark matter.

Persons of Note

1. Airy, George (1801–1982). Astronomer after which the Airy pattern is named.
2. Alvarez, Luis (1911–1988). Lead scientist that proposed that a large asteroid impacting Earth caused the geological K-T/K-Pg boundary.
3. Aristotle (384–322 BC). Influential Greek philosopher largely credited as being the first scientist.
4. Aristarchus (ca. 310–230 BC). Greek astronomer that first determined the distance to the moon, attempted to determine the distance to the sun, and first proposed the sun as the center of the solar system.
5. Bayer, Johann (1572–1625). Responsible for the Bayer system of star names (a Greek letter followed by the constellation name).
6. Brahe, Tycho (1546–1601). Astronomer that made accurate and precise observations of planetary motion that Kepler then went on to use to confirm the heliocentric model.
7. Chandrasekhar, Subrahmanyan (1910–1995). Calculated an accurate maximum mass for a white dwarf star now known as the Chandrasekhar limit.
8. Cook, James (1728–1779). Explorer whose first voyage had the mission of observing the 1769 transit of Venus needed to establish an accurate distance to the sun.
9. Copernicus, Nicolaus (1473–1543). Astronomer that re-established the heliocentric model first proposed by Aristarchus.
10. Crabtree, William (1610–1644). One of the first astronomers to observe a Venus transit for the purposes of establishing the scale of the solar system.
11. Democritus (ca. 460–370 BC). Greek philosopher who first suggested atomic theory in tandem with Leucippus.

12. Doppler, Christian (1803–1853). Physicist that proposed the Doppler shift in frequency of a source observed to be moving toward or away from an observer.
13. Eratosthenes (ca. 276–195 BC). Greek astronomer that calculated the size of Earth.
14. Fermi, Enrico (1901–1954). Influential physicist renowned for his work on nuclear physics and for a simple reasoning technique for quickly solving complex problems.
15. Fowler, Ralph (1889–1944). Physicist and astronomer known for his work on white dwarf stars. Doctoral advisor to Chandrasekhar and Paul Dirac (among many others).
16. Fraunhofer, Joseph (1787–1826). Optician and discoverer of dark absorption bands in the solar spectrum now known as Fraunhofer lines.
17. Galilei, Galileo (1564–1642). Physicist and astronomer to first apply a telescope to astronomy. Discoverer of Jupiter's four large Galilean moons.
18. Goodricke, John (1764–1786). Astronomer known for his observations of variable stars.
19. Halley, Edmund (1656–1742). Astronomer known for successfully predicting the reappearance of a comet.
20. Heisenberg, Werner (1901–1976). Physicist that proposed the uncertainty principle, one of the founding components of quantum mechanics.
21. Hipparchus (ca. 190–120 BC). Greek astronomer known for calculating trigonometric tables and for establishing the magnitude system for star brightness.
22. Hohmann, Wolfgang (1880–1945). Engineer who proposed the elliptical Hohmann transfer orbit between two planets.
23. Horrocks, Jeremiah (1618–1641). One of the first astronomers to observe a Venus transit for the purposes of establishing the scale of the solar system.

24. Hubble, Edwin (1889–1953). Astronomer who established an appearance-based galaxy classification scheme and discovered the expansion of our universe.
25. Joule, James Prescott (1818–1889). Physicist after which the joule unit of energy is named.
26. Kepler, Johannes (1571–1630). Determined the orbits of planets are elliptical, and that went on to establish the founding principles of orbital dynamics.
27. Kuiper, Gerard (1905–1973). Astronomer best known for proposing the source of short-period comets now known as the Kuiper Belt.
28. Leavitt, Henrietta Swan (1868–1921). Astronomer that discovered the period–luminosity relation in Cepheid variables that has been used to establish the distance scale in our universe that is the foundation of cosmology.
29. Leucippus (ca. 490–420 BC). Greek philosopher who first suggested atomic theory in tandem with Democritus.
30. Malmquist, Gunnar (1893–1982). Astronomer who first realized the Malmquist bias introduced by the struggle to detect fainter and fainter objects at greater and greater distances.
31. Newton, Isaac (1642–1726). Influential physicist that played a central role in the enlightenment. Responsible for the foundations of modern physics, including classical mechanics and an accurate model of gravitation.
32. Ockham, William (1285–1347). Franciscan friar known for the Occam's razor approach to reasoning.
33. Oort, Jan (1900–1992). Astronomer best known for proposing the source of long-period comets now known as the Oort Cloud.
34. Oppenheimer, Julius Robert (1904–1967). Physicist most well known for his role in developing nuclear weapons that use processes similar to those in stars.

35. Pauli, Wolfgang (1900–1958). Physicist that discovered the Pauli principle that excludes, for example, two electrons from having the same quantum state at the same time.
36. Pigott, Edward (1753–1825). Astronomer known for his observations of variable stars.
37. Ptolemy, Claudius (ca. 100–168). Roman astronomer who proposed the Ptolemaic geocentric model that included planetary epicycles (circles within a circle).
38. Pythagoras (ca. 570–495 BC). Greek mathematician known for his proof relating the lengths of right-angled triangles.
39. Watt, James (1736–1819). Engineer after which the watt unit of power is named.

Notes

Notes from the Preface
1. Balley, M. "Applied Astronomy Saves the World," *Physics World* 6(2) (1993): 22.
2. Lu, E. Astronomy Saves the World: Protecting the Planet from Asteroid Impacts. Talk given at the University of Hawaii, Manoa, August 2013. The opening slide to the talk lists Protecting the Planet from Asteroids or How Astronomy Saves the World as the title, but the event was advertised using the first title. A video of this lecture can be found online at:
https://connect.arc.nasa.gov/p6riri56zmi/.

Notes from Chapter 1
1. The *New Horizons* mission to Pluto launched from Cape Canaveral Air Force Station on January 19, 2006. It encountered Pluto on July 14, 2015.
http://pluto.jhuapl.edu.
2. Pollack, J. "A Nongrey Calculation of the Runaway Greenhouse: Implications for Venus' Past and Present," *Icarus* 14(3) (1971): 295-501.
3. *MythBusters* was a science and engineering TV show that aired between 2003 and 2016 on The Discovery Channel. http://www.discovery.com/tv-shows/mythbusters/.
4. The *Hayabusa* mission launched from the Uchinoura Space Center in Japan on May 9, 2003. It touched down on Asteroid 25143 Itokawa on November 19, 2005. It returned to Earth on June 13, 2010 where a capsule containing particles from the asteroid was recovered.
5. The *Rosetta* mission launched from the Centre Spatial Guyanais (Guiana Space Centre) on March 2, 2004. Its *Philae* lander touched down on Comet 67P/Churyumov–

Gerasimenko on November 12, 2014. As of this writing, the spacecraft is still in orbit of the comet.
6. *Pioneer 10* and *11* launched from Cape Canaveral Air Force Station on March 2, 1972 and April 6, 1973. Pioneer 10 was the first mission to visit Jupiter. Pioneer 11 also visited Jupiter and then went on to be the first mission to Saturn.
7. *Voyager 1* and *2* launched from Cape Canaveral Air Force Station on September 5, 1977 and August 20, 1977. *Voyager 1* is now in interstellar space and is the farthest human-made object from Earth. *Voyager 2* is the only mission to have visited Uranus and Neptune.
8. Cohen, H. *The Scientific Revolution* (Chicago: The University of Chicago Press, 1994).
9. The Center for the Study of Existential Risk considers how the advancement of technology could result in the extinction of our species. See http://cser.org.
10. See "AC 61-107A – Operations of Aircraft at Altitudes above 25,000 Feet MSL and/or Mach Numbers (MMO) Greater Than 0.75," Table 1.1. Department of Transportation, Federal Aviation Administration.
11. Ade, P. A. R., et al. "Detection of B-Mode Polarization at Degree Angular Scales by BICEP2," The BICEP2 Collaboration, *Phys. Rev. Lett.* 112 (2014): 241101.
12. Cowen, R. "Telescope Captures View of Gravitational Waves," *Nature News*, Nature Publishing Group, March 17, 2014.
13. Ade, P. A. R., et. al. "Joint Analysis of BICEP2/Keck Array and Planck Data," The BICEP2/Keck and Planck Collaborations, *Phys. Rev. Lett.* 114 (2015): 101301.
14. Riley, J. C. "Estimates of Regional and Global Life Expectancy, 1800-2001," *Population and Development Review* 31(3) (2005): 537-543.
15. "Failure is not an option" is not the mantra of any space agency. It is a line from a movie (*Apollo 13*). The scientific

method ensures that failure is always an option. In fact, it can be the preferred outcome for research.
16. Best, B. P. "Nuclear DNA Damage as a Direct Cause of Aging," *Rejuvenation Research* 12 (2009): 199-208.
17. On Feb 26, 2015, Senator James Inhofe (R-OK), chairman of the Environment and Public Works Committee, brought a snowball onto the floor of the U.S. senate as evidence that climate change science is a hoax. https://www.youtube.com/watch?v=3E0a_60PMR8
18. Jervis, R. "Controversial Texas Textbooks Headed to Classrooms," *USA TODAY*, November 17, 2014.
19. Sagan, C. *The Demon-Haunted World: Science as a Candle in the Dark* (Random House, Ballantine Books, 1995).
20. Goitom, H. "Laws Criminalizing Apostasy in Selected Jurisdictions," *The Law Library of Congress*, Global Legal Research Center (2014).
21. Pew Research Center, America's Changing Religious Landscape, May 12, 2015.

Notes from Chapter 2
1. Renne, P. R. et al. "Time Scales of Critical Events around the Cretaceous-Paleogene Boundary," *Science* 339 (2013): 684-687.
2. Jablonski, D., Chaloner, W. G. "Extinctions in the Fossil Record (and Discussion)," *Philosophical Transactions of the Royal Society of London, Series B*. 344 (1994): 11-17.
3. Raup, D., Sepkoski, J. "Mass Extinctions in the Marine Fossil Record." *Science* 215 (1982): 1501-1503.
4. Alvarez, L. W., Alvarez, W., Asaro. F., Michel, H. V. "Extraterrestrial Cause for the Cretaceous–Tertiary Extinction," *Science* 208 (1980): 1095-1108.
5. Haxel, G. B., Hedrick, J. B., Orris, G. J. "Rare Earth Elements—Critical Resources for High Technology," *U.S. Geological Survey*, Fact Sheet 087-02 (2002).

6. Chalk, found below the K-T boundary, is a compression of ancient, tiny, marine species that thrive in shallow, warm oceans. See, for example, Moheimani, N. R., Webb, J. P., Borowitzka, M. A. "Bioremediation and Other Potential Applications of Coccolithophorid Algae: A Review," *Algal Research* 1 (2012): 120-133.
7. Hildebrand, A. R., et al. "Chicxulub Crater: A Possible Cretaceous/Tertiary Boundary Impact Crater on the Yucatán Peninsula, Mexico," *Geology* 19(9) (1991): 867.
8. Newton, I. *Philosophiae Naturalis Principia Mathematica*, Londini, Jussu Societatis Regiæ ac Typis Josephi Streater. Prostat apud plures Bibliopolas. (1687).
9. Motte, A. *The Mathematical Principles of Natural Philosophy, by Sir Isaac Newton*, London, Middle-Temple-Gate Fleet Street (1729). The punctuation used by Motte is replicated here, hence the unusual use of capital letters and apostrophes. See, for example, https://books.google.com/books?id=Tm0FAAAAQAAJ, book 1, pages 19-20.
10. This is not the method Einstein used to derive the $E = mc^2$ equation. This is just a demonstration of dimensional analysis. $E = mc^2$ comes from considering kinetic energy and special relativity.
11. The Peekskill meteorite is a historic object for more than simply hitting Michelle Knapp's car. The event happened on a Friday evening on October 9, 1992. This is football season, and around the time camcorders were just becoming popular. As a result, there are sixteen different videos of the meteor before it made contact with the car. This allows us to trace the origin of the meteor back to an elliptical orbit that passed into the inner edge of the main asteroid belt. See Beech, M., et al. "The Fall of the Peekskill Meteorite: Video Observations, Atmospheric Path,

Fragmentation, Record and Orbit," *Earth, Moon and Planets* 68 (1995): 189-97.

12. The original Beyer-Melosh impact code was hosted at the Planetary Image Research Lab at the University of Arizona: http://pirlwww.lpl.arizona.edu/~jmelosh/crater_c.cg. An updated version is now hosted at http://impact.ese.ic.ac.uk/ImpactEffects/. An impact simulator using the lasted code is available at http://www.purdue.edu/impactearth/

13. Collins, G. S., Melosh, H. J., Marcus, R. A. "Earth Impact Effects Program: A Web-based Computer Program for Calculating the Regional Environmental Consequences of a Meteoroid Impact on Earth," *Meteoritics & Planetary Science* 40(6) (2005): 817-840.

14. NASA's *Mars Reconnaissance Orbiter* spotted a new crater on Mars between March 27 and 28, 2012. Before and after images were serendipitously taken with the Mars Color Imager weather camera.

15. On June 30, 1908, a massive explosion of a meteor occurred several miles above central Russia near the Tunguska River. Despite this meteor being about one hundred meters across, there were no known human casualties (as this area is barely populated) and no crater. Over eight hundred square miles of trees were flattened in a radial pattern. If this event had happened over a major city, the damage would have been catastrophic.

16. On February 15, 2003, another massive explosion occurred several miles above central Russia over the city of Chelyabinsk. This twenty-meter meteor caused widespread damage, including broken windows and a collapsed roof, estimated at over thirty million dollar.

17. The Barringer/Meteor crater is between Flagstaff and Winslow in northern Arizona. A fifty-meter metallic

meteorite, large fragments of which have been recovered, made this 1.2-kilometer crater about fifty thousand years ago. If this type of impact occurred at or near a city today, the death toll could be in the millions.
18. The Wolfe Creek crater is about sixty-five miles from the nearest town in Western Australia. That nearest town, Halls Creek, would almost fit inside the eight-hundred-meter crater likely formed by a meteor a few meters across.
19. Weissman, P. R., et al. "Galileo NIMS Direct Observation of the Shoemaker–Levy 9 Fireballs and Fall Back," *Abstracts of the Lunar and Planetary Science Conference* 26 (1995): 1483.
20. Solem, J. C. "Cometary Breakup Calculations Based on a Gravitationally-Bound Agglomeration Model: The Density and Size of Comet Shoemaker-Levy 9," *Astronomy and Astrophysics* 302 (1995): 596-608.
21. *Deep Impact* was launched on January 12, 2005 from Cape Canaveral Air Force Station and arrived at Comet 9P/Tempel on July 3, 2005. The probe impacted the comet July 4, 2005 at a speed of 23,000 mph.
22. The *Solar and Heliospheric Observatory* was launched on December 2, 1995 from Cape Canaveral Air Force Station. It has been observing the sun for over twenty years and has spotted over three thousand comets.
23. Raddi, R., et al. "Likely Detection of Water-Rich Asteroid Debris in a Metal-Polluted White Dwarf," *Monthly Notices of the Royal Astronomical Society* 450 (2015): 2083-2093.
24. Moore, W. A Treatise on the Motion of Rockets. To Which is Added, An Essay on Naval Gunnery. (London: G. and S. Robinson, Military Academy at Woolwich, 1813).
25. Shoemaker, E., Helin, E. "Earth-Approaching Asteroids as Targets for Exploration," *Asteroids: An Exploration Assessment* , NASA Conference Publication 2053 (1978): 245-256.

26. NEO delta-v catalog: http://echo.jpl.nasa.gov/~lance/delta_v/delta_v.rendezvous.html
27. Altwegg, K., et al. "67P/Churyumov-Gerasimenko, a Jupiter Family Comet with a High D/H Ratio," *Science* 347 (2015): 6220.
28. The *Hayabusa* 2 mission launched from the Tanegashima Space Center in Japan on December 3, 2014. It's expected to arrive at Asteroid 162173 Ryugu in the summer of 2018 and return to Earth in late 2019.
29. NASA's *OSIRIS-REx* mission launched from Cape Canaveral Air Force Station on September 8, 2016. Its mission to Asteroid Bennu should have a significant sample returned by 2023.
30. https://www.nasa.gov/mission_pages/asteroids/initiative/
31. For NEO search programs, see http://neo.jpl.nasa.gov/programs/
32. https://www.zooniverse.org
33. https://www.lsst.org
34. http://neo.jpl.nasa.gov/risks/
35. Giorgini, J. D., et al. "Predicting the Earth Encounters of (99942) Apophis," *Icarus* 193 (2008): 1-19.

Notes from Chapter 3
1. The three relations are known as Charles's law, Boyle's law, and Gay-Lussac's law. They are the fundamental equations of thermodynamics and can be manipulated into the *combined gas law*.
2. Asplund, M., Grevesse, N., Sauval, A. J., Scott, P. "The Chemical Composition of the Sun," *Annual Review of Astronomy & Astrophysics* 47(1) (2009): 481-522.

3. Bellini, G., et al. "Precision Measurement of the Be7 Solar Neutrino Interaction Rate in Borexino," *Physical Review Letters* 107 (2011): 141302.
4. Kearns, E., Kajita, T., Totsuka, Y. "Detecting Massive Neutrinos," *Scientific American* 281(2) (1999): 48-55.
5. Leighton, R. B. "The Solar Granulation," *Annual Review of Astronomy and Astrophysics* 1 (1963): 19.
6. Joy, A. H. "Radial Velocities of Cepheid Variable Stars," *Astrophysical Journal* 86 (1937): 363.
7. Woosley, S. E., Heger, A., Weaver, T. A. "The Evolution and Explosion of Massive Stars," *Reviews of Modern Physics* 74(4) (2002): 1015-1071.
8. Mazzali, P. A., et al. "A Common Explosion Mechanism for Type Ia Supernovae," *Science* 315(5813) (2007): 825.

Notes from Chapter 4
1. The *Alpha Magnetic Spectrometer* was launched to the International Space Station on May 16, 2011 using Space Shuttle *Endeavour*. Its mission is to detect antimatter and dark matter in the space environment. See http://ams.nasa.gov.
2. Hubble, E. P. *The Realm of the Nebulae*. (Yale University Press, 1936).
3. The celestial sphere is an imaginary surface onto which appear all objects in our universe and under which Earth rotates. Coordinate systems are projected onto this imaginary sphere for the purpose of cataloging astronomical objects.
4. The Pythagorean proof is surprisingly simple and can be demonstrated using the areas of three squares. Take two squares (a and b), one bigger than the other, and arrange them such that two corners are touching but the sides are all parallel or perpendicular. The side of a third square (c) can be made by joining the two nearest corners from the

first two blocks. The area of the two original blocks added together equals the area of the third square, i.e., $a^2 + b^2 = c^2$.
5. The distance to the sun is 150,000,000 kilometers.
6. The distance to the Andromeda galaxy is 24,000,000,000,000,000,000 kilometers.
7. Flat-Earthers are people who appear to genuinely believe Earth is flat. It is actually difficult for me to believe that members of the Flat Earth Society are not simply practical jokers. For someone to believe that Earth is flat, everyone else on Earth, including other members of their own society, would need to be carrying out a hoax. One wonders what would happen if you put a flat-Earther in New York, and one in LA, got the two to phone one another at 8 am EST, and then asked them to have a conversation about the altitude of the sun at each location.
8. The Vehicle Assembly Building is 160 meters tall, 218 meters long, and 158 meters wide.
9. The towers of the Golden Gate Bridge are 230 meters tall and spaced 1,280 meters apart.
10. To an ancient Greek like Eratosthenes, Aswan would have been known as Syene.
11. If s is the length of the sagitta, l the length of half of the arc base, and r the radius of the circle (in this case, the size of Earth's shadow at the distance of the moon), then we can find that $r = s/2 + l^2/2s$.

Notes from Chapter 5
1. http://www.politifact.com
2. http://www.snopes.com
3. http://www.hoax-slayer.com
4. http://lmgtfy.com
5. In academic circles, fact-finding is given the more grandiose term "literature review." Before scientific

journals were available online, this could have taken months or years to do a thorough job. Now, even though this exercise must not be rushed, an experienced researcher may be able to get that same task done in a few weeks.
6. Haub, C. *How Many People Have Ever Lived on Earth?* Population Reference Bureau, 2011.
7. https://www.fhwa.dot.gov/ohim/onh00/bar8.htm
8. Venus can be bright enough to see during the day, especially with a properly equipped telescope. Jupiter can also be picked up through careful daytime telescope observations. Any daytime observations must pay careful attention to the position of the sun, and they very rarely have any scientific worth.
9. Belmonte, J. A. "The Decans and the Ancient Egyptian Skylore: An Astronomer's Approach," *Memorie della Società Astronomica Italiana* 73(1) (2002): 43.
10. This is even clearer when you consider the names of the week in French or Spanish.

Notes from Chapter 6
1. Munro, N. D. "Small Game, the Younger Dryas, and the Transition to Agriculture in the Southern Levant," *Mitteilungen der Gesellschaft für Urgeschichte* 12 (2003): 47.
2. Hopkin, M. (2005-02-16). "Ethiopia is Top Choice for Cradle of Homo Sapiens," *Nature News,* February 16, 2005.
3. McPherron, S. P., et al. "Evidence for Stone-Tool-Assisted Consumption of Animal Tissues before 3.39 Million Years Ago at Dikika, Ethiopia," *Nature* 466 (2010): 857-860.
4. Rogers, J. J. W., Santosh, M. *Continents and Supercontinents* (Oxford: Oxford University Press, 2004), 146.
5. Schirrmeister, B. E., et al. "Evolution of Multicellularity Coincided with Increased Diversification of Cyanobacteria and the Great Oxidation Event,"

Proceedings of the National Academy of Sciences 110(5) (2012): 1791-1796.
6. Bell, E. A., et al. "Potentially Biogenic Carbon Preserved in a 4.1 Billion-Year-Old Zircon," *Proceedings of the National Academy of Sciences* 112(47) (2015): 14518-14521.
7. *Hipparcos* was launched on August 8, 1989 from the Centre Spatial Guyanais (Guiana Space Centre). It operated in a highly elliptical Earth orbit until 1993.
8. There is a well-known legend that after the Battle of Agincourt in 1415, English and Welsh archers used the V sign to taunt the defeated French. Supposedly, if captured, the French would remove those two fingers from known enemy archers.
9. The luminosity of the sun is 380,000,000,000,000,000,000,000,000 watts.
10. Between the eighth and thirteenth centuries, significant advances in mathematics and science were made. The nucleus for this Islamic Golden Age was the city of Baghdad.
11. http://www.iau.org/public/themes/constellations/
12. Leavitt, H. S., Pickering, E. C. "Periods of 25 Variable Stars in the Small Magellanic Cloud," *Harvard College Observatory Circular* 173 (1912): 1-3.
13. Hubble, E. P. "Cepheids in Spiral Nebulae," *Popular Astronomy* 33 (1925): 252-255.
14. Until 1949, the 2.5-meter-diameter Hooker telescope was the largest telescope in the world. It is located at the Mount Wilson Observatory in the Angeles National Forest, less than twenty miles from downtown Los Angeles.
15. Hubble, E. P. "A Relation between Distance and Radial Velocity among Extra-Galactic Nebulae," *Proceedings of the National Academy of Sciences* 15(3) (1929): 168-173.

16. Gratton, R. G., et al. "Distances and Ages of NGC 6397, NGC 6752 and 47 Tuc," *Astronomy and Astrophysics* 408 (2003): 529-543.
17. Schroder, K-P., Robert C. Smith, R. C., "Distant Future of the Sun and Earth Revisited," *Monthly Notices of the Royal Astronomical Society* 386 (2008): 155-163.
18. Daniel Batcheldor, *Ice Melting Radiometric Decay Analog, version 1*, [Video file] retrieved from https://www.youtube.com/watch?v=C2gXwOF83E8 (September 12, 2016).
19. Daniel Batcheldor, *Ice Melting Radiometric Decay Analog, version 2*, [Video file] retrieved from https://www.youtube.com/watch?v=6cqhOnlV0pc (September 12, 2016).
20. Bouvier, A., Wadhwa, M. "The Age of the Solar System redefined by the Oldest Pb–Pb Age of a Meteoritic Inclusion," *Nature Geoscience*, 3, (2010)637 - 641.

Notes from Chapter 7
1. Krauss, L. A Universe from Nothing: Why There Is Something Rather than Nothing (Free Press, 2012).
2. Perlmutter, S., Schmidt, B. P., Riess, A. G. The Nobel Prize in Physics 2011 "for the discovery of the accelerating expansion of the Universe through observations of distant supernovae."
3. Carroll, S. M. "The Cosmological Constant" *Living Reviews in Relativity* 4 (2001): 1.
4. Binney, J., Tremaine, S. *Galactic Dynamics* (Princeton University Press, 1987).
5. Rubin, V. C., Ford, W. K., Jr. "Rotation of the Andromeda Nebula from a Spectroscopic Survey of Emission Regions," *Astrophysical Journal* 159 (1970): 379.
6. Rubin, V. C., Thonnard, N., Ford, W. K., Jr., Roberts, M. S. "Motion of the Galaxy and the Local Group Determined

from the Velocity Anisotropy of Distant Sc I Galaxies. II. The Analysis for the Motion," *Astronomical Journal* 81 (1976): 719.
7. Rubin, V. C., Ford, W. K. Jr., Thonnard, N. "Rotational Properties of 21 Sc Galaxies with a Large Range of Luminosities and Radii from NGC 4605 (R=4kpc) to UGC 2885 (R=122kpc)," *Astrophysical Journal* 238 (1980): 471.
8. http://wwwmpa.mpa-garching.mpg.de/galform/virgo/millennium/
9. http://wwwmpa.mpa-garching.mpg.de/gadget/
10. http://hipacc.ucsc.edu/Bolshoi/
11. Kashlinsky, A. "LIGO Gravitational Wave Detection, Primordial Black Holes, and the Near-IR Cosmic Infrared Background Anisotropies," *The Astrophysical Journal Letters*, 823(2) (2016): 25.
12. Milgrom, M. "A Modification of the Newtonian Dynamics as a Possible Alternative to the Hidden Mass Hypothesis," *Astrophysical Journal* 270 (1983): 365-370.
13. Bergeron, P., Ruiz, M. T., Leggett, S. K. "The Chemical Evolution of Cool White Dwarfs and the Age of the Local Galactic Disk," *The Astrophysical Journal Supplement Series* 108(1) (1997): 339-387.

Notes from Chapter 8
1. Cameron, A. G. W. "The Origin of the Moon and the Single Impact Hypothesis V," *Icarus* 126(1) (1997): 126-137.
2. Plait, P. Death from the Skies! The Science behind the End of the World (Penguin Books Ltd, 2008).
3. *Kepler* was launched from Cape Canaveral Air Force Station on March 7, 2009. In 2012, *Kepler* lost the ability to precisely point itself but carried on for almost another year gathering useful data. http://kepler.nasa.gov
4. *Pale Red Dot* is a team of astronomers looking for evidence of any planets around Proxima Centauri. Their latest

results, announced in August 2016, show that there is some good, but not conclusive, evidence of a planet around Proxima that may have a mass similar to that of Earth. See https://palereddot.org and Anglada-Escudé, G., et al. "A terrestrial planet candidate in a temperate orbit around Proxima Centauri" *Nature* 536 (2016): 437-440.
5. http://breakthroughinitiatives.org/Initiative/3
6. Apollo 1 was to be the first crewed flight of the Apollo program with the ultimate mission of sending humans to the moon. During training, the three astronauts, Gus Grissom, Ed White, and Roger Chaffee, tragically lost their lives in a fire inside the cockpit of the command module.
7. Heavy-lift launch vehicles (HLLVs) are those that can carry more than twenty tons into low-earth orbit. Future proposed HLLVs include the SpaceX Falcon Heavy (http://www.spacex.com/falcon-heavy)
and NASA Space Launch System
(https://www.nasa.gov/exploration/systems/sls/index.html).
8. Margot, J. L., et al. "Topography of the Lunar Poles from Radar Interferometry: A Survey of Cold Trap Locations," *Science* 284(5420) (1999): 1658.
9. http://www.nasa.gov/mission_pages/LCROSS/main/prelim_water_results.html
10. https://www.nasa.gov/resource-prospector
11. Shapiro, J. R., Schneider, V. "Countermeasure Development: Future Research Targets," *Journal of Gravitational Physiology* 7 (2000): 1.
12. LeBlanc, A. et al. "T2 Vertebral Bone Marrow Changes After Space Flight," *Magnetic Resonance in Medicine* 41 (1999): 495-498.
13. http://www.planetaryresources.com/
14. https://deepspaceindustries.com

15. Hohmann, W. *Die Erreichbarkeit der Himmelskörper*. Verlag Oldenbourg in München (1925).
16. http://er.jsc.nasa.gov/seh/ricetalk.htm
17. Excerpt from Edward Young's (1683–1765) *Night-Thoughts* poem.

Notes from Chapter 9
1. United Nations, "Committee on the Elimination of Discrimination against Women," *Convention on the Elimination of All Forms of Discrimination against Women*, CEDAW/C/AFG/1-2, Article 10 (2011).
2. International Humanist and Ethical Union, Freedom of Thought 2012: A Global Report on Discrimination against Humanists, Atheists and the Non-religious (2012). http://www.iheu.org/files/IHEU Freedom of Thought 2012.pdf
3. Chopra, D. Quantum Healing: Exploring the Frontiers of Mind/Body Medicine (Bantam, 1989).
4. TEDMED, *Deepak Chopra at TEDMED 2010*, [Video file] retrieved from https://www.youtube.com/watch?v=tglicKfBpc0 (March 7, 2011, 3:15 and 3:35).
5. Strickler, L. *Senate Panel Probes 6 Top Televangelists*, CBS News Investigative Unit, November 6, 2007.
6. The non-profit Trinity Foundation tracks religious fraud and provides information on these practices and help for victims: http://trinityfi.org
7. Stenger, V. "Quantum Quackery," *Skeptical Inquirer*, 21.1 (January/February 1997).
8. The *Urban Dictionary* defines a Chopraism as "A string of common words placed within a sentence so incomprehensible the listener is sometimes rendered temporarily comatose."

http://www.urbandictionary.com/define.php?term=Chopraism
9. Bernardo, S. *Richard Dawkins Interviews Deepak Chopra (Enemies of Reason Uncut Interviews)*, [Video file] retrieved from https://www.youtube.com/watch?v=qsH1U7zSp7k (May 4, 2013, 2:20).
10. Petigura, E. A., Andrew W. Howard, A. W., Geoffrey W. Marcy, G. W. "Prevalence of Earth-size Planets Orbiting Sun-like Stars," *Proceedings of the National Academy of Sciences* 110(48) (2013): 19273-19278.
11. US Congress, "Appropriations Committee Releases the Fiscal Year 2012 Commerce, Justice, Science Appropriations," *US House of Representatives Committee on Appropriations* (July 6, 2011). http://appropriations.house.gov/news/documentsingle.aspx?DocumentID=250023
12. Otto, S. L. "Antiscience Beliefs Jeopardize U.S. Democracy," *Scientific American,* November 1, 2012.
13. Kluger, J. "Earthrise on Christmas Eve: The Picture That Changed the World," *Time,* December 24, 2013.
14. McCarthy, M. "Earthrise: The image That Changed Our View of the Planet," *The Independent,* June 12, 2012.
15. The Deep Space Climate Observatory launched from Cape Canaveral Air Force Station on February 11, 2015. It sits at a gravitational null point that allows it to continually face the sunlit side of Earth.
16. Sagan, C. *Pale Blue Dot: A Vision of the Human Future in Space* (Ballantine Books, 1994).
17. Extract from Thomas Jefferson's letter to Pierre Samuel DuPont de Nemours (April 24, 1816).
18. Rich, M. "Nationwide Test Shows Dip in Students' Math Abilities," *The New York Times,* October 28, 2015.

19. Robinson, K. Out of Our Minds: Learning to be Creative (Capstone, 2011).
20. 107th Congress. "No Child Left Behind Act of 2001" *Public Law 107–110* (January 8, 2002) http://www2.ed.gov/policy/elsec/leg/esea02/107-110.pdf

Index

accuracy, 94, 109, 110, 114, 118, 227
Airy pattern, 181, 227, 237
Airy, George, 181
Alexandria, 95
Alpha Centauri, 138, 182, 183
Alpha Magnetic Spectrometer, 166, 248
Alpha Orionis, 140
Alvarez, Luis, 30
Anders, Bill, 217
Andromeda, 90, 91, 95, 133, 142, 147, 150, 227, 249, 252
anti-intellectualism, 17, 227
Apollo 17, 185, 217
Apollo 8, 217
apophenia, 115, 227
Apophis, 54, 247
Aristarchus of Samos, 101
Aristotle, 117, 237
artificial gravity, 187, 194, 198
Asteroid Robotic Redirect Mission, 50, 189
Asteroid Zoo, 53
astronomical unit, 101, 103, 120, 121, 122, 124, 125, 126, 127, 128, 134, 138, 167, 168, 169, 182, 195, 200, 201, 218, 227, 231, 232, 233
Aswan, 95, 115, 249

B2 Spirit, 191
baloney, 21, 24, 25
Barringer (crater), 40, 245
Bayer, Johann, 139, 237
Bennu, 50, 190, 247

Berkshire Hathaway, 192
Betelgeuse, 139, 140
Beyer, Ross A., 39
BICEP2, 16, 17, 242
big bang, 23, 25, 150, 151, 160, 161, 166, 227, 228
binary stars, 72, 73, 107, 168, 169, 227
black hole, 8, 59, 60, 72, 166, 174, 177, 198, 228, 229
Blue Marble, 218, 219
blueshift, 144, 228
Bolshoi Simulation, 165
Borman, Frank, 217
Bouvier, Audrey, 156, 252
Brahe, Tycho, 118, 120, 122, 123, 237
Buffett, Warren, 192

celestial sphere, 84, 248
Center for the Study of Existential Risk, 14, 242
Cepheid variables, 139, 140, 141, 142, 145, 147, 149, 228, 239, 248
Cepheus, 140
Ceres, 200
Chandrasekhar, Subrahmanyan, 74, 173, 174, 228, 235, 237, 238
Chelyabinsk, 40, 245
Chicxulub (crater), 31, 39, 40, 244
chondrite, 154, 228
Chopra, Deepak, 208, 209, 255, 256

Churyumov-Gerasimenko (67P), 10, 43, 44, 48, 49, 190, 241, 247
citizen science, 52
CNO cycle, 68, 228
COLBERT, 188
Colbert Report, 188
Colbert, Stephen, 188
Colombia, 153
congress, 215, 243, 256, 257
convection zone, 67
Cook, James, 125, 237
Copernicus, Nicolaus, 117, 118, 237
cosine, 86, 87, 88, 89, 228, 235
cosmic microwave background, 16, 151, 228
cosmic ray, 80, 228
Crabtree, William, 125, 237
Cretaceous, 30, 31, 228, 244

dark energy, 161, 212, 229
dark matter, 80, 83, 145, 164, 165, 166, 212, 229, 232, 236, 248
Deep Field, 217
Deep Impact, 43, 246
Deep Space Climate Observatory, 218
degenerate, 174
Deimos, 195, 199
Delta Cephei, 140, 141, 149
delta-v, 46, 47, 48, 49, 54, 187, 192, 195, 229, 234, 247
Democritus, 61, 237, 239
di Bondone, Giotto, 43
dimensional analysis, 35, 36, 131, 148, 149, 168, 229, 244
Doppler effect, 143, 144, 145, 229

Doppler, Christian, 143
Drake equation, 213, 214
Drake, Frank, 213

Earthrise, 217, 218, 219, 256
eccentricity, 120
electrolysis, 45
electron degeneracy pressure, 174, 229
equilibrium, 66, 68, 70, 71, 73, 74, 159, 160, 161, 162, 164, 172, 173, 174, 229, 230
Eratosthenes, 94, 100, 238, 249
ESA, 42, 132
Eta Aquilae, 140
event horizon, 60, 229
extraterrestrial, 214, 219
Exxon Mobile, 192

Fermi, Enrico, 103, 104, 109, 229, 238
fission, 60, 229
flux, 137, 138, 229
Forbes, 192
Fowler, Ralph, 173, 238
Fraunhofer, Joseph, 144, 238
fuel, 8, 44, 45, 46, 47, 48, 51, 66, 67, 70, 154, 190, 192, 194, 195, 196, 197, 234
fusion, 60, 64, 65, 66, 67, 68, 71, 73, 74, 144, 228, 229

GADGET, 165
Galaxy Zoo, 53
Galilei, Galileo, 113, 118, 142, 238, 246
geoid, 23, 25, 109
Giotto, 42

globular clusters, 151, 152, 171, 172, 229
Goldilocks, 104, 230
Goodricke, John, 140, 238
granules, 67, 230
gravitational lensing, 165, 166, 230
gravitational wave, 16, 80, 230
gravity assist, 48, 49, 230

half-life, 155, 156, 228, 230
Halley, Edmund, 42, 126
Hawking, Stephen, 183
Hayabusa, 10, 49, 50, 190, 241, 247
heat shield, 195, 196
Heisenberg uncertainty principle, 173, 230
Heisenberg, Werner, 173
heliocentric, 116, 117, 118, 119, 120, 122, 237
Hipparchus of Nicaea, 87, 133
Hipparcos, 133, 138, 139, 149, 251
HMS Queen Elizabeth, 191
Hoba (meteorite), 38
Hohmann, Wolfgang, 193, 238, 255
Homo erectus, 184
Hooker telescope, 142, 251
Horrock, Jeremiah, 125, 238
Hubble Space Telescope, 80, 95, 132, 149, 217, 221
Hubble, Edwin, 142
huckster, 208, 209
hypergolic, 196, 230

impulse, 35

In situ resource utilization, 197, 199
inferior planet, 116, 121, 122, 123, 124
infinity, 88, 161
inflation, 16, 190, 230
International Space Station (ISS), 47, 48, 95, 166, 167, 185, 186, 188, 189, 190, 191, 198
Internet, 104, 105, 165, 216
inverse square law, 136, 137, 138, 231
iridium, 30, 31
isotope, 63, 67, 68, 154, 155, 156, 157, 230, 231, 232, 234
Itokawa, 10, 49, 190, 241

James Webb Space Telescope, 215, 219
Jefferson, Thomas, 220, 256
Joule, James Prescott, 47
JP Morgan Chase, 192

Kelly, Mark, 188
Kelly, Scott, 188
Kennedy, John F., 204
Kepler Space Telescope, 182
Kepler, Johannes (Space Telescope), 118, 119, 120, 123, 124, 125, 168, 169, 182, 193, 231, 237, 239, 253
kinetic energy, 36, 37, 159, 171, 231, 236, 244
Knapp, Michelle, 37, 244
K-T/K-Pg boundary, 41, 237, 244
Kuiper Belt, 43, 200, 201, 228, 231, 239

Kuiper, Gerard, 43

Lake Acraman (crater), 40
Large Synoptic Survey Telescope, 53
Leavitt, Henrietta Swan, 141, 142, 148, 239, 251
Leucippus, 61, 237, 239
logarithm, 106, 231
logical fallacy, 20, 21, 22, 23, 231
Lovell, Jim, 217
low-Earth orbit, 47, 48, 179, 180, 187, 190, 192, 195, 196, 203
luminosity, 136, 137, 138, 140, 141, 228, 231, 251
lunar eclipse, 96, 97, 98, 99, 114, 206, 231
Lunar Resource Prospector, 186

Magellanic Clouds, 90, 141, 147, 231, 235, 251
main sequence, 69, 70, 71, 73, 152, 172, 173, 231
Malmquist bias, 134, 139, 231, 239
Malmquist, Gunnar, 134
mass spectrometer, 156, 232
massive stars, 67, 68, 70, 72, 74, 151, 154, 172
Melosh, H. Jay, 39
Milky Way, 82, 133, 150, 152, 231, 232, 235
Millennium Simulation Project, 165
Milner, Yuri, 183
Mir, 188
moon base, 185, 187, 188, 190, 191, 195

Motte, Andrew, 31
MU-69, 201
MythBusters, 9, 241

NASA, 11, 23, 43, 48, 49, 51, 186, 189, 190, 218, 245, 246, 247, 254
Near-Earth Objects, 51, 52, 53, 54, 55, 56, 247
neutron degeneracy pressure, 174, 232
neutron star, 8, 72, 172, 174, 232
New Horizons, 9, 183, 241
Newton, Isaac, 31, 32, 34, 124, 168, 239, 244
No Child Left Behind, 223, 257
nothing, 14, 19, 34, 52, 59, 60, 71, 75, 156, 160, 161, 162, 170, 210, 213, 219, 252

occultation, 55
Ockham, William of, 15
Oort Cloud, 43, 44, 201, 202, 228, 232, 239
Oort, Jan, 43, 201
Oppenheimer, Julius Robert, 172, 239
order of magnitude, 106, 108, 113, 163, 193, 232
Orion Nebula, 90, 170
OSIRIS-REx, 49, 50, 190, 247

Pale Blue Dot, 218, 219, 256
Paleogene, 31
Pangaea, 153
parallax, 111, 112, 113, 114, 124, 127, 128, 129, 132, 133, 141, 149, 169, 233

parsec, 129, 163, 233
Pauli exclusion principle, 173, 233
Pauli, Wolfgang, 173
peer-review, 22
Philae, 43, 49, 241
Phobos, 195, 199
Pigott, Edward, 140, 240
Pillars of Creation, 217
Pioneer 10 and 11, 11, 242
Planck constant, 143, 233, 242
planetary nebula, 75, 233
planetary science, 8, 9, 27, 245, 246
Pluto, 9, 44, 183, 200, 201, 218, 231, 241
point spread function, 181
Polyakov, Valeri, 188
post-theism, 26, 233
potential energy, 159, 168, 171, 236
precision, 90, 98, 99, 102, 109, 110, 112, 115, 118, 120, 122, 125, 132, 133, 157, 173, 233
Principia, 244
Procyon, 138
Proxima Centauri, 128, 129, 182, 233, 253
Ptolemaic, 117, 118, 233, 240
Ptolemy, Claudius, 117, 240
Pythagoras, 89, 240

quantum mechanics, 1, 10, 161, 173, 174, 209, 211, 234, 236, 238

radian, 85, 86, 87, 88, 90, 91, 93, 126, 128, 234, 235
radiative zone, 67
radioactive decay, 155, 228, 236
radiometric dating, 132, 154, 157, 232, 234
random walk, 65, 69
red giant, 74
redshift, 144, 145, 234
Rees, Martin, 14
refraction, 132, 234
relaxation timescale, 163, 164, 171, 234
retrograde, 116, 117, 206, 234
Rigel, 139
Robinson, Ken, 223
rocket equation, 46, 234
Rosetta, 10, 43, 48, 49, 50, 190, 241
Rubin, Vera, 164, 252, 253
Ryugu, 49, 247

Sagan, Carl, 11, 21, 218, 243, 256
sagitta, 98, 234, 249
Saturn V, 48, 185
scientific method, 15, 16, 17, 20, 22, 65, 151, 159, 166, 207, 211, 212, 221, 223, 234, 243
Seager equation, 214
Seager, Sara, 214
sectarian, 19
Severny, 60
Shackleton crater, 186
Shoemaker-Levy 9, 41, 44, 217, 246
significant figures, 109, 156, 234
sine, 87, 88, 89, 90, 121, 234, 235
Sirius, 73, 128, 129, 138, 139, 234
small angle approximation, 90, 235

Solar and Heliospheric Observatory, 43, 246
solar eclipse, 55, 95, 96, 114, 233, 236
spectral energy distribution, 142, 143, 144, 235
spectrograph, 143, 235
speed of light, 37, 73, 127, 143, 144, 174, 228, 229
spin-off, 13
Sputnik, 177
standard candle, 139, 140, 141, 142, 151
star formation, 151, 170
STEAM, 80, 235
STEM, 8, 10, 80, 88, 89, 220, 223, 235
strong force, 63, 235
Sudbury (crater), 40
sunrise, 27, 94
sunset, 27
superior planet, 117, 122, 123
supernova, 72, 74, 153, 154, 174, 228, 235, 248, 252
synodic, 123, 125, 193, 235

tangent, 88, 90, 93, 113, 235
Tau Ceti, 138
Televangelists, 209, 255
Tempel (9P), 43, 44, 246
Tequila, 71
Theia, 178
thermonuclear, 60, 65, 66, 68, 70, 71, 73, 144, 153, 173
Torino scale, 54, 177, 235
transit spectroscopy, 219, 236
trigonometry, 86, 87, 88, 93, 95, 102, 113, 121, 122, 126, 222
Tsar Bomba, 60, 61
tsunamis, 39
Tunguska, 40, 245

unit circle, 86, 87, 88, 89, 236
USS Gerald Ford, 191

variable stars, 70, 140, 141, 142, 236, 238, 240
Vega 1 and 2, 42
virial theorem, 171, 236
Voyager 1 and 2, 11, 218, 242
Vredefort (crater), 40

Wadhwa, Meenakshi, 156, 252
water, 29, 38, 43, 44, 45, 46, 47, 49, 50, 51, 67, 69, 71, 94, 104, 105, 107, 108, 132, 155, 180, 184, 185, 186, 194, 195, 197, 199, 200, 201, 213, 214, 230, 254
Watt, James, 136
weak force, 62, 236
white dwarf, 8, 73, 74, 75, 172, 173, 174, 228, 229, 233, 235, 236, 237, 238
WIMPs, 166, 236
Wolfe Creek (crater), 40, 246

Yucatán Peninsula, 31, 39, 244

Zooniverse, 52, 53
Zuckerberg, Mark, 183